DUST CONTROL HANDBOOK

DUST CONTROL HANDBOOK

by

Vinit Mody and Raj Jakhete

Martin Marietta Laboratories
Baltimore, Maryland

NOYES DATA CORPORATION
Park Ridge, New Jersey, U.S.A.

Copyright © 1988 by Noyes Data Corporation
Library of Congress Catalog Card Number 88-19061
ISBN: 0-8155-1182-5
ISSN: 0090-516X
Printed in the United States

Published in the United States of America by
Noyes Data Corporation
Mill Road, Park Ridge, New Jersey 07656

10 9 8 7 6 5 4 3 2 1

Library of Congress Cataloging-in-Publication Data

Mody, V.
        Dust control handbook / by V. Mody and R. Jakhete.
            p. cm. -- (Pollution technology review, ISSN 0090-516X ; no.
    161)
        Bibliography: p.
        Includes index.
        ISBN 0-8155-1182-5 :
        1. Metallurgical plants--Dust control--Handbooks, manuals, etc.
    I. Jakhete, R. II. Title. III. Series.
    TH7697.M4M63   1988
    669--dc19   88-19061
                                CIP

Foreword

This handbook consolidates information developed by industry and government laboratories on dust control engineering techniques for metal and nonmetal mineral processing. Although designed for the minerals processing industry, the technology is pertinent for other industries as well.

The concepts of dust, its prevention, formation and control are examined, including wet and dry control systems, personal protection, and testing methods. Costing methodologies are also examined.

Prospective users of this handbook include maintenance foremen, plant engineers, mill supervisors, and plant safety directors, or the like. The information is described and illustrated in a manner to assist one to have enough information to implement a dust control technique without need to refer to other sources.

The information in the book is from *Dust Control Handbook for Minerals Processing*, prepared by V. Mody and R. Jakhete of Martin Marietta Laboratories for the U.S. Department of Interior, Bureau of Mines, February, 1987.

NOTICE

The material in this book was prepared as an account of work sponsored by the U.S. Department of Interior, Bureau of Mines. On this basis neither Martin Marietta nor the Publisher assumes any responsibility or liability for the accuracy, adequacy, or completeness of the concepts, methodologies, or protocols described in this handbook.

This handbook is intended only as a guide. Dust controls for individual sites must be designed, implemented, and operated according to applicable federal, state, and local regulations and specific site conditions.

Final determination of the suitability of any information or product for use contemplated by any user, and the manner of that use, is the sole responsibility of the user. The book is intended for informational purposes only. The reader is warned that caution must always be exercised when dealing with potentially hazardous materials or processes such as those involved in dust control, and expert advice should be obtained before implementation is considered.

Contents

Chapter 1
Dust and Its Control

What Is Dust?

Dust consists of tiny solid particles carried by air currents. These particles are formed by a disintegration or fracture process, such as grinding, crushing, or impact. The Mine Safety and Health Administration (MSHA) defines dust as finely divided solids that may become airborne from the original state without any chemical or physical change other than fracture.

A wide range of particle sizes is produced during a dust generating process. Particles that are too large to remain airborne settle while others remain in the air indefinitely.

Dust is generally measured in micrometers (commonly known as microns). Some common objects and their size in microns are listed below.

	μm
Red blood corpuscles	8
Human hair	50-75
Cotton fiber	15-30

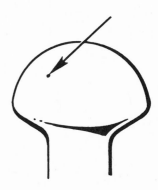

A Micron-Size Dust Particle on a Pin Head

How Is Dust Generated?

Dust is generated by a wide range of manufacturing, domestic, and industrial activities. Construction, agriculture, and mining are among the industries that contribute most to atmospheric dust levels.

In minerals processing operations, dust is emitted—

- When ore is broken by impact, abrasion, crushing, grinding, etc.

1

Photo of Respirable Dust

● Through release of previously generated dust during operations such as loading, dumping, and transferring

● Through recirculation of previously generated dust by wind or by the movement of workers and machinery

The amount of dust emitted by these activities depends on the physical characteristics of the material and the way in which the material is handled.

Types of Dust

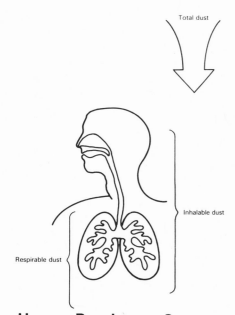

Human Respiratory System

Fibrogenic dust, such as free crystalline silica (FCS) or asbestos, is biologically toxic and, if retained in the lungs, can form scar tissue and impair the lungs' ability to function properly.

Nuisance dust, or inert dust, can be defined as dust that contains less than 1% quartz. Because of its low content of silicates, nuisance dust has a long history of having little adverse effect on the lungs. Any reaction that may occur from nuisance dust is potentially reversible. However, excessive concentrations of nuisance dust in the workplace may reduce visibility (e.g., iron oxide), may cause unpleasant deposits in eyes, ears, and nasal passages (e.g., portland cement dust), and may cause injury to the skin or mucous membranes by chemical or mechanical action.

From an occupational health point of view, dust is classified by size into three primary categories:

● Respirable dust

● Inhalable dust

● Total dust

Respirable Dust

Respirable dust refers to those dust particles that are small enough to penetrate the nose and upper respiratory system and deep into the lungs. Particles that penetrate deep into the respiratory system are generally beyond the body's natural clearance mechanisms of cilia and mucous and are more likely to be retained.

MSHA defines respirable dust as the fraction of airborne dust that passes a size-selecting device, having the following characteristics:

Aerodynamic diameter, μm (unit density spheres)	Percent passing selector
2.0	90
2.5	75
3.5	50
5.0	25
10.0	0

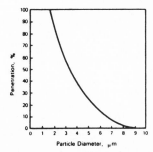

Definition of Respirable Dust

Inhalable Dust

The EPA describes inhalable dust as that size fraction of dust which enters the body, but is trapped in the nose, throat, and upper respiratory tract. The median aerodynamic diameter of this dust is about 10 μm.

Total dust includes all airborne particles, regardless of their size or composition.

Total Dust

Why Is Dust Control Necessary?

Although unavoidable in many minerals processing operations, the escape of dust particles into the workplace atmosphere is undesirable. Excessive dust emissions can cause both health and industrial problems:

- Health hazards
  - Occupational respiratory diseases
  - Irritation to eyes, ears, nose, and throat
  - Irritation to skin

- Risk of dust explosions and fire

- Damage to equipment

- Impaired visibility

- Unpleasant odors

- Problems in community relations

Of greatest concern is the health hazard to workers who are excessively exposed to harmful dusts. In order to evaluate the severity of health hazard in a workplace, the American Conference of Governmental

Industrial Hygienists (ACGIH) has adopted a number of standards, commonly known as threshold limit values (TLV's). These values are used as guides in the evaluation of health hazards. TLV's are time-weighted concentrations to which nearly all workers may be exposed 8 hours per day over extended periods of time without adverse effects. MSHA uses these TLV's for health hazard evaluation and enforcement.

Health Hazard Factors

Not all dusts produce the same degree of health hazard; their harmfulness depends on the following factors:

- Dust composition
  - Chemical
  - Mineralogical

- Dust concentration
  - On a weight basis: milligrams of dust per cubic meter of air (mg/m^3)
  - On a quantity basis: million particles per cubic foot of air (mppcf)

- Particle size and shape
  - The particulate size distribution within the respirable range
  - Fiberous or spherical

- Exposure time

Excessive or long-term exposure to harmful respirable dusts may result in a respiratory disease called pneumoconiosis. This disease is caused by the buildup of mineral or metallic dust particles in the lungs and the tissue reaction to their presence. Pneumoconiosis is a general name for a number of dust-related lung diseases. Some types of pneumoconiosis are:

Healthy Lung   Contaminated Lung

- **Silicosis** — Silicosis is a form of pneumoconiosis caused by the dust of quartz and other silicates. The condition of the lungs is marked by nodular fibrosis (scarring of the lung tissue), resulting in shortness of breath. Silicosis is an irreversible disease; advanced stages are progressive even if the individual is removed from the exposure.

- **Black Lung** — Black lung is a form of pneumoconiosis in which respirable coal dust particles accumulate in the lungs and darken the tissue. This disease is progressive. Although this disease is commonly known as black lung, its official name is coal worker's pneumoconiosis (CWP).

- **Asbestosis** — Asbestosis is a form of pneumoconiosis caused by asbestos fibers. This disease is also irreversible.

How Is Dust Controlled?

Dust control is the science of reducing harmful dust emissions by applying sound engineering principles. Properly designed, maintained, and operated dust control systems can reduce dust emissions and, thus, workers' exposure to harmful dusts. Dust control systems can also reduce equipment wear, maintenance, and downtime; increase visibility; and boost employee morale and productivity.

Reducing employee exposure to dust can be accomplished by three major steps:

- Prevention
- Control systems
- Dilution or isolation

Prevention

The saying "prevention is better than cure" can certainly be applied to the control of dust. Although total prevention of dust in the bulk material handling operation is an impossible task, properly designed bulk material handling components can play an important role in reducing dust generation, emission, and dispersion.

Control Systems

After all the necessary preventive measures have been adopted, the dust still remaining in the workplace can be controlled by one or more of the following techniques: dust collection systems, wet dust suppression systems, and airborne dust capture through water sprays.

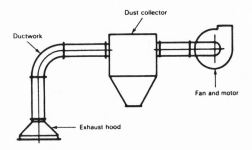

Dust Collection System

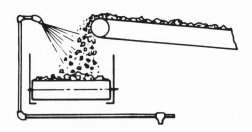

Wet Dust Suppression System

Dilution - Isolation

Dust Collection Systems

Dust collection systems use industrial ventilation principles to capture airborne dust from the source. The captured dust is then transported to a dust collector, which cleans the dusty air.

Wet Dust Suppression Systems

Wet dust suppression systems use liquids (usually water) to wet the material so that it has a lower tendency to generate dust. Keeping the material damp immobilizes the dust, and very little material becomes airborne.

Airborne Dust Capture Through Water Sprays

This technique suppresses airborne dust by spraying fine droplets of water on the dust cloud. The water droplets and dust particles collide and form agglomerates. Once these agglomerates become too heavy to remain airborne, they settle from the airstream.

Dilution Ventilation

This technique reduces the dust concentration in the area by diluting the contaminated air with uncontaminated fresh air. In general, dilution ventilation is not as satisfactory for health hazard control or dust collecting systems; however, it may be applied in circumstances where the operation or process prohibits other dust control measures.

Isolation

Isolation is another means to protect workers from exposure to harmful dust. In this technique, the worker is placed in an enclosed cab and supplied with fresh, clean, filtered air.

Chapter 2
Preventing Dust Formation

In any minerals processing facility, dust is generated when ore is shattered or broken as in dumping, loading, transferring, or handling. Proper design, selection, and operation of equipment to minimize ore breakage can therefore reduce dust.

Two primary groups of equipment are used in minerals processing operations:

- Processing equipment such as crushers, screens, grinding mills, and dryers

- Bulk material handling equipment such as belt conveyors, screw conveyors, bucket elevators, feeders, and hoppers

Processing equipment processes the ore into a final product. Bulk material handling equipment transfers ore between processing equipment and from the mine to the processing facility.

Processing equipment must be retrofitted with dust control measures; dust control measures can be implemented when bulk material handling equipment is designed. Although the choice of equipment for a specific operation is based on process needs, the use of alternate equipment, improved equipment design, or sometimes even a change in process can greatly reduce dust emissions. Simple measures such as providing shrouds, covers, or enclosures around a dust source can also help to contain dust emissions or allow the existing dust control system to operate more efficiently.

7

The following sections describe the most commonly used processing and bulk material handling equipment, their major dust emission points, and measures to prevent or reduce dust generation, emission, and dispersion.   Several common minerals processing operations are also discussed.

Belt Conveyors

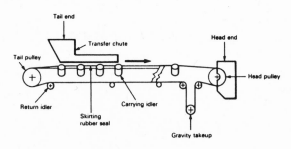

Belt Conveyor

The belt conveyor is one of the major pieces of equipment used to transfer ore between processing equipment and from one point to another within a minerals processing facility. It is also one of the most frequent sources of dust emissions.

Types of Belt Conveyors

Troughed Belt

Flat Belt

There are two primary types of belt conveyors:

- Troughed belt conveyors
- Flat belt conveyors

Troughed belt conveyors are the most common type of conveyor used in mining and minerals processing operations.   Flat belt conveyors are used mostly in nonmining industries.

Emission Points

Belt conveyors emit dust from the following four points:

- The tail end, where material is received
- The conveyor skirting

● The return idlers, due to carryback of fine dust on the return belt

● The head end, where material is discharged

The following measures should be considered in the design or selection of a belt conveyor.

Belt Loading

The amount of dust generated at belt conveyor transfer points depends on the way the material is loaded onto the belt. To reduce dust generation—

● The material should be loaded onto the center of the belt.

● The material and the belt should travel in the same direction and at the same speed, whenever possible.

Impact at Loading Point

A momentary deflection of the belt between two adjacent idlers may result when ore strikes the belt. As a result, a puff of dust may leak out under the skirting rubber seal. To prevent dust emissions at the loading point, adequately spaced impact idlers (1-ft centers) should be located at transfer points. These will absorb the force of impact and prevent deflection of the belt between the idlers, thus preventing dust leakage under the skirting rubber seal.

Conveyor Skirting

Skirtboards are used to keep the material on the belt after it leaves the loading chute. They are equipped with flat rubber strips that provide a dust seal between the skirtboards and the moving belt.

The conventional skirtboard design uses vertical rubber strips. This design is not recommended for the following reasons:

● The vertical rubber seals wear out quickly.

● The rubber must be adjusted constantly to prevent dust leakage, and this is often neglected.

Dust Prevention Measures

Momentary Deflection of Belt

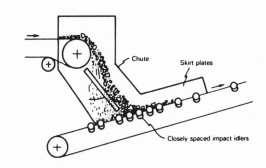

Impact Idler Location

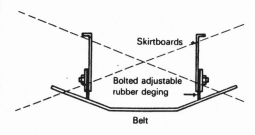

NOT RECOMMENDED

Conventional Skirting

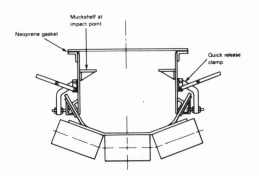

Improved Skirting

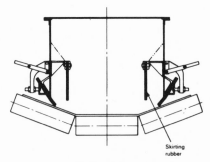

Double Skirting

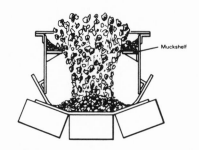

Muckshelf

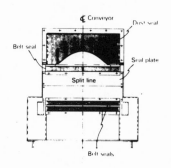

Dust Seal at Head End

Illustrated is an improved skirting design, which has the following important features:

● The skirtboards are sufficiently high and wide to accommodate both the volume of material and the pressure surges caused by the inflowing material and induced air.

● Quick-disconnect clamps are used instead of conventional bolts for fast, easy adjustment of the rubber.

● The flexibility of the inclined skirting rubber allows it to rest on the moving belt at all times, even when the belt is momentarily deflected between the idlers.

● The inclined skirting rubber is 1/2 in. thick with 60-65 durometer hardness and provides a greater wear area for increased life.

● The top edges of the skirtboards are covered and sealed with self-adhesive neoprene rubber gaskets for a proper dust seal.

A double skirting can be installed in the impact zone if the incoming material hits the skirting rubber directly, increasing the wear rate.

Muckshelves

Muckshelves can be installed in the belt conveyor's material impact zone to—

● Load the material centrally on the belt and keep the belt properly aligned

● Protect the inclined skirting rubber from direct impact with the incoming material

Dust Curtains

Dust curtains are used to contain dust within a conveyor enclosure. They should be installed at the head, tail, and exit ends. Dust curtains are made of rubber with 60-65 durometer hardness and can be hinged at the conveyor's head and exit ends to provide easy access during maintenance.

Belt Cleaners

Belt Scrapers — A belt scraper should be installed at the head pulley to dislodge fine dust particles that may adhere to the belt surface and to reduce carryback of fine materials on the return belt. A scrapings chute should also be provided to redirect the material removed by the belt scraper into the process stream or container.

V-Plow — Product spillage or dust leakage may fall on the noncarrying side of the belt and eventually build up on the surfaces of the tail pulley. This buildup may move the belt laterally and thus make the skirting rubber seals ineffective. A V-plow installed on the noncarrying side of the belt will clean the belt and prevent buildup of material and dust on the tail pulley, thus keeping the belt properly aligned.

Conveyor Capacity

The belt conveyor should be designed to operate at 75% of its full rated capacity. This reduces spillage, dust emission, and wear on skirting rubber seals.

The following measures are suggested to adjust the loading capacity of existing conveyors:

- Increase the belt speed

- Change the idlers' angle (for example, from 20° to 35°)

- Increase the conveyor width (for example, from 24 to 36 in.)

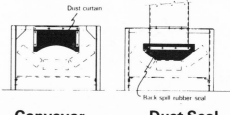

Conveyor Dust Curtain   **Dust Seal at Tail End**

b = Belt Width (in.)	A	B	C
24	19	18	6
30	24	21	8
36	29	27	9
42	36	32	11
48	41	36	12

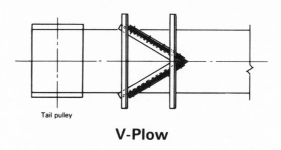

V-Plow

Transfer Chutes

Transfer chutes transport ore from one piece of equipment to another. Significant dust generation can result if the transfer chute is not designed properly.

Dust Prevention Measures

The following points should be considered when designing a transfer chute:

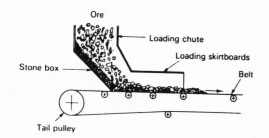

Rockbox

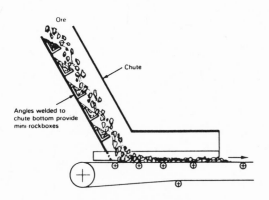

Mini-Rockboxes

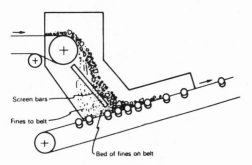

**Fines and Lumps
on Belt**

 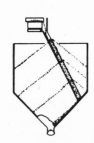

Spiral Chute   Bin-Lowering Chute

● The chute should be big enough to avoid jamming of material and reduce air flow. When fine material and lumps are mixed in the product stream, the chute depth should be at least three times the maximum lump size to avoid jamming.

● The chute should be designed so the material falls on the sloping bottom of the chute and not on the succeeding equipment.

● Wherever possible, the material should fall on a local rockbox or stonebox rather than on the metal surfaces to—
 - Reduce dust and noise generation
 - Absorb the impact of incoming material
 - Reduce wear and abrasion of the chute surfaces
 - Reduce the height of material fall
 - Reduce dust emission from the backspill rubber seal at the tail end of the conveyor

Note: When handling fine or abrasive materials, a number of small steel angles can be welded on the chute bottom to form mini-rockboxes. The oncoming material slides on the material stored in the rockboxes, greatly reducing wear and abrasion of the chute bottom.

● Abrupt changes of direction must be avoided to reduce the possibility of material buildup, material jamming, and dust generation.

● Curved, perforated, or grizzly chute bottoms should be used when the product stream consists of fines and lumps. Placing a layer of fines ahead of the lumps on the belt helps prevent heavy impact of material on the belt, which reduces belt wear and dust generation.

Other types of chutes are used for controlling dust during bulk material handling:

● Spiral chutes are used to prevent breakage of fragile or soft material.

● Bin-lowering chutes are used to feed bins and hoppers without generating large amounts of dust. These chutes consist of a channel that runs from the discharging equipment down into the bin. The channel is secured to the sloping side of the bin. Ore slides down the chute quietly, with minimum dusting. When the material meets the bin side or the surface of the material in the bin, it leaves the chute from the sides and spreads out conically.

- Rock ladders are used to prevent breaking and crumbling of rock. They consist of a steel tower with a series of baffles (or mini-rockboxes) arranged so the discharged rock never has a free drop of more than 5 to 6 ft. This lower drop height significantly reduces dust.

- Telescopic chutes are used to minimize the height of material fall into stockpiles. The telescopic sections are usually cable-connected so a winch can lift the sections of the chute. The lower end of the chute is always kept just clear of the top of the stockpile to reduce dust.

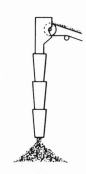

Telescopic Chute

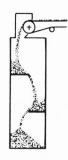

Rock Ladder

Enclosures are used to contain dust emissions around a dust source. They can also lower the exhaust volume requirements of a dust collection system or help make the existing dust collection system work more efficiently.

Enclosures

The following guidelines are suggested when designing an enclosure for a dust source:

- Enclosures should be spacious enough to permit internal circulation of the dust-laden air.

- Enclosures should be arranged in removable sections for easy maintenance.

- A hinged access door should be provided to aid routine inspection and maintenance.

- Dust curtains should be installed at the open ends of the enclosures to contain dust and reduce air flow.

Dust Prevention Measures

Exhausted enclosures

(a)   (b)

Narrow Enclosure   **Spacious Enclosure**

Crushers

Crushers reduce coarse material to a desired size. The crushing process uses mechanical energy and rubbing to fracture the rock. The forces applied to rock fragments during crushing processes are—

- Compression force
- Impact force

Compression is a slow application of force on the rock while impact is a short, sudden application of force.

All crushers generate dust. Crushers that primarily use impact forces produce large amounts of fines and dust. Those that primarily use compression forces produce dust in proportion to the stage of reduction: dust production increases progressively from first- to third-stage crushing.

Types of Crushers

There are six main categories of crushers used in minerals processing operations:

- Jaw crushers
- Gyratory crushers
- Cone crushers
- Hammermills
- Impact breakers
- Roll crushers

Jaw, gyratory, and cone crushers primarily use compression forces; hammermills, impact breakers, and roll crushers use impact forces.

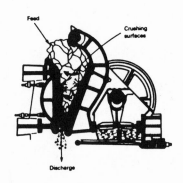

Jaw Crusher

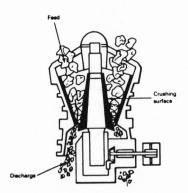

Gyratory Crusher

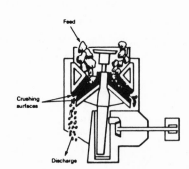

Cone Crusher

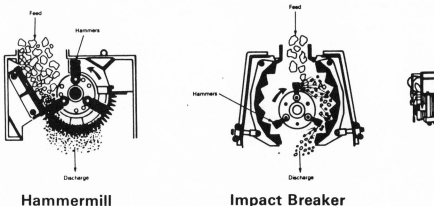

Hammermill   **Impact Breaker**   **Double-Roll Crusher**

Emission Points

Crushers emit dust primarily from two points:

- Crusher discharge
- Crusher feed

Dust Prevention Measures

Dust control measures are not usually considered in the design of a crusher. However, the use of shrouds or enclosures for crushers can contain the dust so that a dust control system can operate more efficiently. The following measures are recommended:

- A crusher feedbox with a minimum number of openings should be installed. Rubber curtains should be used to minimize dust escape and air flow.

- The crusher should be choke fed to reduce air entrainment and dust emission.

Dust escape at the crusher discharge end can be minimized by properly designed and installed transfer chutes.

Screens

Screens are used to sort material according to size. The material fed into a screen is separated into at least two sizes:

- Undersize material, which passes through the screen opening

- Oversize material, which is retained on the screen surface

Although screening can be either wet or dry, dry screening is most often used in minerals processing operations.

Dust is generated in all dry screening processes. However, the amount of dust depends on the particle size contained in the ore, the moisture content, and the type of screening equipment used. Generally, a screen processing finer material produces more dust. Also, screens agitated harder and faster produce more dust than those vibrated more gently and slowly.

Types of Screens

The four most common types of screening equipment are—

- Grizzlies
- Shaking screens
- Vibrating screens
- Revolving screens

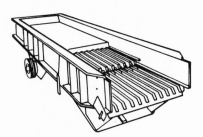

Grizzly Screen

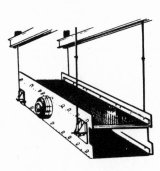

Shaking Screen

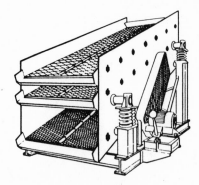

Vibrating Screen

Revolving Screen

Emission Points

In general, screens emit dust from the following points:

- The top one-third of the screen surface where incoming material hits

- The openings between moving parts (the screen) and stationary parts (the discharge chutes)

- Discharge chutes

Dust Prevention Measures

The rate of dust generated by screens cannot be altered. However, properly enclosing the screen can reduce dust emissions. A complete enclosure that can be easily removed for maintenance and inspection should be used. One commercially available enclosure system consists of a special rubber cloth, rubber molding, and simple metal hardware. The rubber cloth can be attached to the screen to provide an almost perfect dust seal between the screen and the discharge chutes. The top of the screen can also be enclosed by the rubber cloth to prevent dust escape. These lightweight rubber covers not only provide a dust-tight enclosure but also allow easy maintenance, inspection, and replacement of screening surfaces.

A tight sealing system reduces dust emissions and also minimizes air flow, which reduces the exhaust volume for the dust collection system installed downstream.

Some screen manufacturers provide sheet-metal covers to enclose the top of the screen. These covers are effective when properly maintained. However, they do not provide a dust seal between the moving screen surfaces and the stationary chutes.

Storage Bins and Hoppers

Bins and hoppers are used to store ore temporarily. They act as buffers to absorb the surge between unloading and consumption.

The ore is fed into the bin by various equipment, such as conveyors, elevators, and screens. The material is normally discharged from bins and hoppers through gravity or vibrating feeders.

Emission Points

Bins and hoppers primarily emit dust from—

- Feed openings
- Discharge feeder
- Inspection doors

Dust Prevention Measures

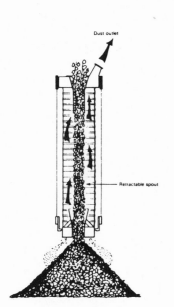

Loading Spout

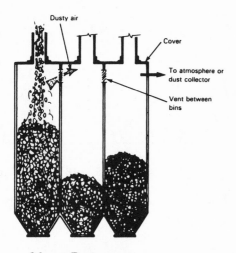

**Vent Between
Bins/Silos**

Dust emissions during feeding operations can be minimized by—

- Installing a bin-lowering chute.

- Completely enclosing the bin or hopper. When bins and hoppers are enclosed completely, an inspection door or a bin-level detector should be installed so the material level can be monitored.

Dust emissions during material discharge can be minimized by—

- Installing a telescopic chute.

- Installing a loading spout. Such spouts are sophisticated versions of the telescopic chute and are used to load and stack ore into barges, trucks, and railroad cars. The spouts apply three basic principles of dust control:

  - Containment
  - Dust capture close to the source
  - Preventing air flow caused by the falling material

The falling material is enclosed by a flexible duct, acting as a chute, which retracts as the height of the material pile increases. The duct also prevents air flow during free fall of material between the chute and stockpile. The generated dust is captured by the same flexible duct and is conveyed, countercurrent to the material flow, to a dust collector.

Another method commonly used to reduce dust emission is to transfer dusty air through a vent into an adjoining bin or silo. Of course, this assumes that multiple bins are used, that the bins are totally enclosed, and that the adjoining bin has room for the displaced air to expand.

A typical bucket elevator consists of a series of buckets mounted on a chain or belt that operates over head and foot wheels. A steel casing usually encloses the entire assembly. The buckets are loaded by scooping up material from the boot (bottom) or by feeding material into the buckets. Material is discharged as the bucket passes over the head wheel.

Bucket Elevators

- Centrifugal discharge
- Positive discharge
- Continuous discharge

Types of Bucket Elevators

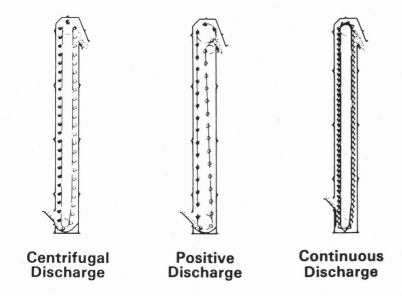

Centrifugal Discharge　　**Positive Discharge**　　**Continuous Discharge**

Bucket elevators emit dust from two points:

- The boot, where material is fed
- The head wheel, where material is discharged

Emission Points

The steel casing that encloses the buckets and chain assembly contains dust effectively unless there are holes or openings in the casing.

Emissions at the boot of the bucket elevator can be reduced by proper design of a transfer chute between the feeding equipment and the elevator. Dust production can be reduced significantly by keeping the

Dust Prevention Measures

height of material fall to a minimum and by gently loading material into the boot of the elevator.

Proper venting to a dust collector, as well as proper enclosures and chutes between the elevator discharge and the receiving equipment, will control dust emission at the discharge end of the bucket elevator.

Feeders

Feeders are relatively short conveyors used to deliver a controlled rate of ore to the processing equipment.

Although dust is emitted from all types of feeders, the amount of dust depends on—

- The kind of material being handled
- The size of material
- The degree of agitation of the material

Types of Feeders

- Apron
- Belt
- Reciprocating
- Vibrating
- Disc

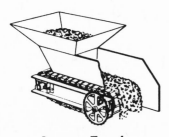

Apron Feeder

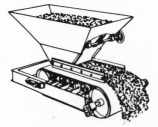

Belt Feeder

Reciprocating Feeder

Vibrating Feeder

Disc Feeder

Dust emission points from a feeder are—

- At the tail end, where material is received
- At the head end, where material is discharged

Reduce dust production during feeding operations by—

- Enclosing feeders as much as possible
- Selecting an oversize feeder or a feeder that produces less agitation of the ore

---

Screw conveyors are one of the oldest and simplest types of equipment used to move ore. They consist of a conveyor screw rotating in a stationary trough. Material placed in the trough is moved along its length by rotation of the screw.

Screw Conveyors

Screw Conveyor

Screw conveyors emit dust primarily from—

- The inlet, where material is received
- Leaks in the trough cover
- Worn-out troughs

Normally, screw conveyors are totally enclosed except at the ends, where emissions can be controlled by proper transfer chute design.

The trough cover is usually fastened by nuts and bolts. However, to maintain a proper dust seal, a self-adhesive neoprene rubber gasket should be installed. Many manufacturers provide two-bar flanges and formed-channel cross members that make a continuous pocket around the trough. The flange-cover sections are set in this channel. Once the channel section is filled with dust, an effective dust seal is created.

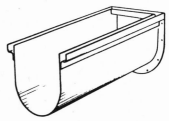

Dust Seal Trough

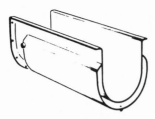

Jacketed Trough

Abrasive materials can wear out screw conveyor troughs quickly unless a special coating or abrasion-resistant material is used for the trough.

Pneumatic Conveyors

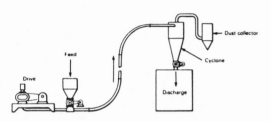

Pressure Conveying System

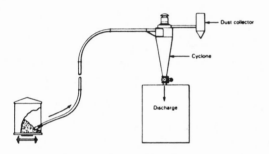

Vacuum (Suction) Conveying System

Pneumatic conveyors are tubes or ducts through which material is moved by pressure or vacuum (suction) systems.

Pressure systems can be either high or low pressure. Low-pressure systems operate at pressures obtainable from a fan; high-pressure systems use a compressed-air source. When material is fed into a pressure system, the airstream immediately suspends it and conveys it to a cyclone- or filter-type collector. The conveying air then escapes through the cyclone vent or a filter.

Vacuum systems offer clean, efficient pickup of material from rail cars, trucks, or bins and hoppers for unloading into other types of equipment. Cyclone receivers or filters are used at the end of such systems to separate the material.

Emission Points

Since pneumatic systems are totally enclosed, dust emissions do not usually occur unless the system has worn-out areas.

Dust Prevention Measures

Because maximum wear in the conveying ductwork occurs at elbows, long radius elbows made of heavy gauge material should be used. The elbows can also be lined with refractory or ceramic material to further reduce the wear and abrasion.

In low-pressure pneumatic systems, dust may leak through joints. Self-adhesive neoprene gaskets should be used at all joints to provide a dust-tight seal.

Grinding and pulverizing reduce ore to a desired fineness for further treatment.

In its basic form, a grinding mill consists of a horizontal, slow-speed, rotating cylindrical drum. Rod, pebble, and ball mills are the most common types of grinding mills used in minerals processing operations. Steel rods, balls, or pebbles roll freely inside the drum during rotation to provide the grinding action.

Grinding Mills

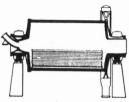

Rod Mill

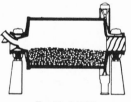

Ball Mill

Dust is emitted from a grinding mill—

- At the inlet, where material is fed
- At the outlet, where material is discharged

Emission Points

Most grinding mills are fed by a belt conveyor or a feeder. The ground ore is discharged to a screen, conveyor, or elevator.

Properly designed enclosures and chutes, as well as rubber dust seals between moving and stationary components, should be installed at the feed and discharge ends to minimize dust emissions.

Dust Prevention Measures

Dryers remove water or other volatile material from solid substances primarily by introducing hot gases into a drying chamber. The hot air readily absorbs moisture from the material.

Rotary dryers, flash dryers, spray dryers, and tray and compartment dryers are a few of the many types of

Dryers

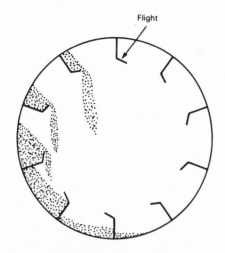

Flight

**Cross-Section View
of a Rotary Dryer**

dryers. However, in minerals processing operations, rotary dryers are the most commonly used.

Rotary dryers consist of a horizontally inclined rotating cylinder. Material is fed at one end and discharged at the other end. In direct-type rotary dryers, hot gases move through the cylinder in direct contact with the material, either with or against the direction of its flow. The cylinder is equipped with flights, which lift the material and shower it down through the hot gas stream. This type of dryer has a potential for high levels of dust emission. In an indirect-type rotary dryer, heat is applied by combustion gases on the outside of the cylinder or through steam tubes inside the cylinder. This type of dryer has much less tendency to emit dust. It is usually used when continuous drying of powdery or fine material is needed.

Emission Points

Dust emission can be a problem in any dryer in which material is agitated or stirred. Dust is emitted only from the discharge end.

A study on rotary dryers by the Barber Greene Company concluded that the dust carryout increased proportionately to the square of the exhaust gas volume.

Dust Prevention Measures

The hot, dust-laden gases from the dryer are carried to a dust collector, and dust normally does not escape unless the ventilation system is worn out or improperly maintained.

Stockpiles

Large volumes of processed material are stored in open or enclosed stockpiles. Open stockpiles are normally used when the material size is large. Enclosed stockpiles are used when material is either very fine or must be stored dry. Stockpiles are considered active when material is continuously removed or added. They are considered inactive when material is not added or withdrawn for long periods. All types of stockpiles can be a significant dust source.

Generation of dust emissions from stockpiles is due to—

- The formation of new stockpiles
- Wind erosion of previously formed piles

During formation of stockpiles by conveyors, dust is generated by wind blowing across the stream of falling material and separating fine from coarse particles. Additional dust is generated when the material hits the stockpile.

Dust from stockpiles can be reduced through the following measures:

- Minimizing height of free fall of material and providing wind protection using—

  - Stone ladders, which consist of a section of vertical pipe into which stone is discharged from the conveyor. At different levels, the pipe has square or rectangular openings through which the material flows to form the stockpile. In addition to reducing the height of free fall of material, stone ladders also provide protection against wind.

  - Telescopic chutes, in which the material is discharged to a retractable chute. As the height of the stockpile increases or decreases, the chute is raised or lowered accordingly. Although some free fall of material from the end of the chute to the top of the stockpile occurs, proper design of the chute can keep the drop to a minimum.

  - Stacker conveyors, which operate on the same principle as telescopic chutes. The conveyor has an adjustable hinged boom that raises or lowers it according to the height of the stockpile.

- Minimizing wind erosion of the stockpile by—
  - Locating stockpiles behind natural or manufactured windbreaks
  - Locating the working area on the leeward side of the active piles
  - Covering inactive piles with tarps or other inexpensive materials

Emission Points

Dust Prevention Measures

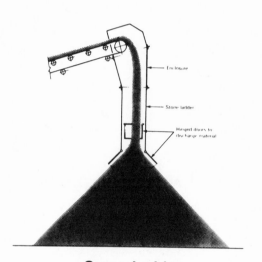

Stone Ladder

- Minimizing vehicle traffic on or around the stockpile

- Using specialized equipment such as a reclaimer to minimize the disturbance of the stockpile or providing a tunnel underneath to reclaim the material

---

Haul Roads

Haul roads are used in minerals processing operations to transport ore from the quarry to the processing plant, as well as within the facility. Large amounts of dust can be generated during this process.

Emission Points

Dust emissions from hauling operations vary, depending on—

- The condition of the road surface
- The volume and speed of vehicle traffic

Dust Prevention Measures

Dust emissions from haul roads can be minimized by—

- Spraying the soil frequently with water, chemicals, oil, or other stabilizing agents

- Paving the haul roads (Note: Paved roads should be cleaned and vacuumed periodically to remove accumulated soil and dust.)

- Reducing traffic volume by replacing small haul vehicles with larger ones

- Reducing and strictly enforcing traffic speed

Ore is dumped or unloaded from trucks or railroad cars in most minerals processing facilities.

Truck and Railroad Car Dumping

Dust generation and emission during the dumping of ore are caused by—

Emission Points

- Dumping large volumes of material in a relatively short time (3 to 10 seconds), which displaces an equal volume of air carrying fine dust particles

- Wind spreading the dust

Enclosures should be used to contain dust during dumping. A sufficiently large enclosure will contain most of the dust, as well as aid internal recirculation of the dust-laden air.

Dust Prevention Measures

Enclosures for railroad-car dumping operations must have openings at each end to allow cars to enter and exit. However, these openings can create a wind-tunnel effect. To minimize this effect, the size of these openings should be kept as small as possible with shrouds, rubber curtains, etc. Shrouds or rubber curtains should also be used for other types of enclosures to reduce the area through which air can escape or enter.

In addition to the above design guidelines, the following operational measures are suggested:

- Increase the dumping cycle time to reduce the rate of displaced air and thus reduce airborne dust emissions.

- Decrease the amount of open area through which the material flows. This measure will reduce the escape of dust through unused areas.

Powder Handling and Packing

Pulverized material, such as sand, silica flour, hydrated lime, or other powdery material, is normally shipped either in bulk quantities through trucks or railroad cars or in small volumes using paper bags, drums, or barrels. Although the process of filling a bag, drum, or barrel is simple, packing fine material can be an extremely dusty operation.

Two types of mini-bulk-packing processes commonly used in minerals processing are—

- Bagging
- Barrel or drum filling

Bagging

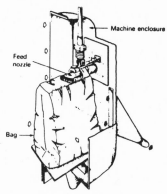

Automatic Bagging Machine
Bureau of Mines

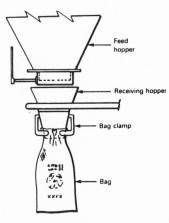

Manual Bagging Machine

This process uses bagging machines commonly known as packers. Several different types of packers, such as fluidized auger or screw type, belt or sling type, drop type, and impeller type, are available. However, the two most commonly used packers in the minerals processing industry are—

- Open-mouth (drop-type) packers using sewn or adhesive-sealed bags

- Spout fluidizing packers used with seal-valve bags

The open-mouth type uses gravity to fill sewn or adhesive-sealed bags. The material is fed into a weigh hopper, and when the correct weight is reached, the feed gate closes and the contents drop into the open-mouth bag.

The spout fluidizing packer uses compressed air to force the material through a nozzle into a paper bag equipped with a seal valve. When the desired weight is reached, the compressed air supply is cut off and the feed stopped. The internal pressure of the contents of the bag then forces the valve to close.

Sometimes the final product is shipped in barrels, drums, or containers. The material is fed into the receptacle by gravity.

Barrel or Drum Filling

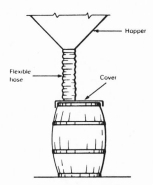

Drum-Filling Operation

Considerable amounts of dust may be emitted during bagging operations. The following are common occurrences:

Emission Points

● **Dust emitted while filling the bag.** Material and dust spillage may occur because of the compressed air used to pressurize the bag.

● **Bag surface dust.** The dust emitted during bagging extremely fine products may stick to the outer surfaces of the bag. This dust may become airborne during later handling, conveying, loading, or stacking of bags. Unfortunately, attempts to clean the bags by air brushing have been unsuccessful.

● **Poor bag quality.** Poor bag quality or improper storage may result in broken bags, leaky valves, or weak bag seams. Leaks from the valve and poorly glued seams may be a continuous source of dust as the bag is moved from the bagging machine to the loading point.

● **Spillage from nozzles.** Spillage from the nozzle of spout fluidizing packers is another major dust source. The bag is usually discharged from the machine immediately after the flow of product into the bag is stopped. However, fluidized material remaining in the nozzle area continues to discharge, which results in spillage. This spillage may also prevent the bag valve from sealing properly, which may cause another dust source as the bag is handled.

Dust generation and emission during barrel- or drum-filling operations result primarily from—

- Displaced air carrying fine dust

- Spillage of material during the filling process and subsequent handling of drums and barrels

Dust Prevention Measures

Dust emissions caused by bagging machines can be minimized in the following ways:

- The bag should be properly attached to the spout to reduce dust leakage or material spillage.

- Where possible, the bag's outer surfaces should be wetted to prevent the surface dust from becoming airborne.

Dust emissions due to poor bag quality can be minimized by—

- Following proper bag specifications

- Eliminating sharp obstructions during bag-handling processes

- Using proper compressed-air pressures to fill the bag, which can significantly reduce the number of broken bags

- Storing bags at recommended temperatures and humidity levels to prevent drying and cracking

In the case of spout fluidizing packers, the following additional measures are suggested:

- The fluidized air in the bag should be vented to control dust emissions during filling. This can be done by using bags with perforations staggered through the layers of bag paper. This eliminates a direct path for the product and reduces dust spillage, while maintaining adequate air relief through the bag. The porosity of the paper further aids air relief and has a filtering effect.

- The nozzle should be cleaned before the bag is released from the machine to prevent spillage of material from the nozzle. Cleaning is accomplished by injecting a short, high-velocity, low-volume blast of compressed air at the rear of the nozzle to fluidize the remaining material in the nozzle and force it into the bag. (This procedure is useful only for fine granular material, such as whole grain sand. For finely ground material such as ground silica, a specialized nozzle developed by the Bureau of Mines may be used.)

Dust generation and emission during drum- or barrel-filling operations can be minimized by—

- Providing a cover with a flexible chute attached to the storage bin

- Enclosing the operation as much as possible to contain the dust

- Reducing the rate of discharge of the material

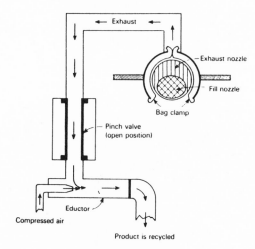

**New USBM
Nozzle System**

Bureau of Mines

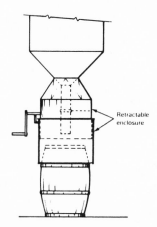

**Dust Enclosure for
Drum-Filling Operation**

Chapter 3
Dust Control Systems

Why Dust Control?

After dust is formed, control systems are used to reduce dust emissions. Although installing a dust control system does not assure total prevention of dust emissions, a well-designed dust control system can protect workers and often provide other benefits, such as—

- Preventing or reducing risk of dust explosion or fire

- Increasing visibility and reducing probability of accidents

- Preventing unpleasant odors

- Reducing cleanup and maintenance costs

- Reducing equipment wear, especially for components such as bearings and pulleys on which fine dust can cause a "grinding" effect and increase wear or abrasion rates

- Increasing worker morale and productivity

- Assuring continuous compliance with existing health regulations

Proper planning, design, installation, operation, and maintenance are essential for an efficient, cost-effective, and reliable dust control system.

Types of Dust Control Systems

The three basic types of dust control systems currently used in minerals processing operations are—

- Dust collection
- Wet dust suppression
- Airborne dust capture

Dust collection systems use ventilation principles to capture the dust-filled airstream and carry it away from the source through ductwork to the collector.

Wet dust suppression techniques use water sprays to wet the material so that it generates less dust.

Airborne dust capture systems may also use a water-spray technique; however, airborne dust particles are sprayed with atomized water. When the dust particles collide with the water droplets, agglomerates are formed. These agglomerates become too heavy to remain airborne and settle.

Selection of a Dust Control System

The selection of a dust control system is normally made based on the desired air quality and existing regulations. Dust collection systems can provide reliable and efficient control over a long period; however, the capital and operating costs are high. Wet dust suppression and airborne dust capture systems, while somewhat less efficient, are less expensive to install and operate but also require careful selection and planning to be most effective.

The facilities that require dust control should be surveyed in detail before a dust control system is selected. Emphasis should be placed on the process, the operating conditions, the characteristics of the processing equipment, associated dust problems, and toxicity of the dust. The following is a list of information that may be required:

- Process flow diagram of the facility indicating items such as the type of material being handled, material flow rates, and the type of equipment

- Major dust emission points and conditions that occur at these points during normal operations

- Desired performance of the system

- Drawings indicating equipment layout

- Retention time of material in bins or stockpiles

- Availability of electrical and other utilities

- Areas requiring freeze protection

The dust collection system, also known as the local exhaust ventilation system, is one of the most effective ways to reduce dust emissions.

A typical dust collection system consists of four major components:

- An exhaust hood to capture dust emissions at the source

- Ductwork to transport the captured dust to a dust collector

- A dust collector to remove the dust from the air

- A fan and motor to provide the necessary exhaust volume and energy

Each of these components plays a vital role in proper operation of a dust collection system, and poor performance of one component can reduce the effectiveness of the other components. Therefore, careful design and selection of each component is critical.

The following sections describe the design of exhaust hoods and ductwork used to remove dust from the process. Chapter 4 discusses dust collectors, fans, and motors, which are used to collect the dust for disposal.

Dust Collection Systems

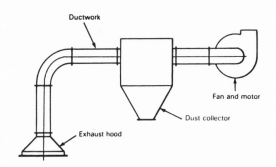

Components of Dust Collection System

Principles of Airflow

Air flows from a high- to a low-pressure zone due to the pressure difference. The quantity and the velocity of airflow are related according to the following equation:

$$Q = AV$$

where:

Q   =   volume of airflow, ft^3/min

V   =   velocity of air, ft/min

A   =   cross-sectional area through which the air flows, ft^2

Air traveling through a duct is acted on simultaneously by two kinds of pressure:

- Static pressure
- Velocity pressure

Both SP and VP are components of a third kind of pressure:

- Total pressure

Static Pressure (SP)

SP = Gauge Pressure – Atmospheric Pressure

Static pressure (SP) is a force that compresses or expands the air. It is used to overcome the frictional resistance of ductwork, as well as the resistance of such obstructions as coils, filters, dust collectors, and elbows.

SP is the difference between pressure in a duct and that in the atmosphere. When the SP is above the atmospheric pressure, it has a positive sign (+); when it is below the atmospheric pressure, it has a negative sign (-). SP is commonly measured in inches of water.

SP always acts perpendicular to the ductwalls and creates outward pressure when positive and inward pressure when negative.

Velocity Pressure (VP)

$$VP = \left(\frac{V}{4005}\right)^2$$

Velocity pressure (VP) is the pressure required to accelerate the air from rest to a particular velocity. It exists only when air is in motion, always acts in the direction of airflow, and is always positive in sign. VP is also commonly measured in inches of water.

Note: The relationship, as illustrated, is valid only when g = 32.2 ft/s^2 (gravitational acceleration constant) and ρ = 0.075 lb/ft^3 (air density). For other conditions, a correction factor must be used.

Total pressure is the algebraic sum of SP and VP. It is the pressure required to start and maintain the air-flow.

If the velocity of air flowing through a duct increases, part of the available SP is used to create the additional VP to accelerate the airflow. Conversely, if the velocity is reduced, a portion of the VP is converted into SP. These conversions, however, are always accompanied by a net loss of TP (in other words, the conversion is always less than 100% efficient).

When air enters a suction opening, the airstream gradually contracts a short distance downstream and, as a result, a portion of the static pressure is converted into velocity pressure. The plane where the diameter of the jet is the smallest is known as the vena contracta. After the vena contracta, the airstream gradually expands to fill the duct and, consequently, a portion of the velocity pressure is converted into static pressure. Both of these pressure conversions are accompanied by losses, which reduce the airflow. The amount of airflow reduction can be defined by a factor known as the coefficient of entry, "C_e."

This represents the percentage of flow that will occur into a given exhaust hood based on the static pressure developed by the hood. It is defined as the actual rate of flow caused by a given static pressure compared to the theoretical flow that would result if there were no losses due to pressure conversions.

Related to the C_e is the term "hood entry loss" or "h_e." It is defined as the factor representing the loss in pressure caused by air flowing into a duct. It is measured in inches of water.

The exhaust hood is the point where dust-filled air enters a dust collection system. Its importance in a dust collection system cannot be overestimated. It must capture dust emissions efficiently to prevent or reduce worker exposure to dusts. The exhaust hood—

- Encloses the dust-producing operation

- Captures dust particulates and guides dust-laden air efficiently

Total Pressure (TP)

$$TP = SP + VP$$

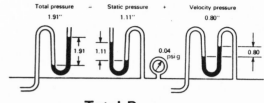

Total Pressure

Vena Contracta

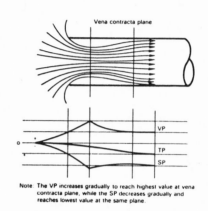

Note: The VP increases gradually to reach highest value at vena contracta plane, while the SP decreases gradually and reaches lowest value at the same plane.

Vena Contracta

"C_e", Coefficient of Entry

$$C_e = \left[\frac{\text{Actual airflow}}{\text{Theoretical airflow}} \right]$$

"h_e", Hood Entry Loss

Exhaust Hood

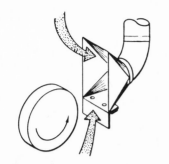

Local Hood

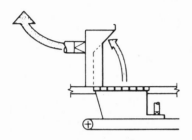

Side Hood

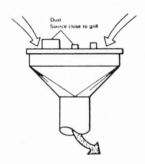

Downdraft Hood

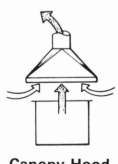

Canopy Hood

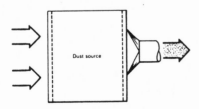

Booth Hood

Types of Exhaust Hoods

The three general classes of exhaust hoods are—

- Local hoods
- Side, downdraft, and canopy hoods
- Booths or enclosures

Local hoods are relatively small structures. They are normally located close to the point of dust generation and capture the dust before it escapes. Local hoods are generally efficient and typically used for processes such as abrasive grinding and woodworking.

Side, downdraft, and canopy hoods are larger versions of local hoods. They also rely on the concept of preventing dust emissions beyond the control zone. They are typically used for plating tank exhausts, foundry shakeouts, melting furnaces, etc. These hoods are generally less efficient than local hoods.

Booth and enclosure hoods isolate the dust-generating process from the workplace and maintain an inward flow of air through all openings to prevent the escape of dust. These hoods are the most popular type in minerals processing operations because they are very efficient at minimum exhaust volumes. They are typically used for areas such as vibrating or rotating screens, belt conveyors, bucket elevators, and storage bins.

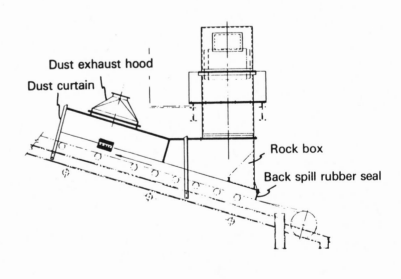

Enclosure Hood

Design of Exhaust Hoods

The design of an exhaust hood requires sufficient knowledge of the process or operation so that the most effective hood or enclosure (one requiring minimum exhaust volumes with desired collection efficiency) can be installed.

The successful design of an exhaust hood depends on—

- Rate of airflow through the hood
- Location of the hood
- Shape of the hood

Of the above three factors, the rate of airflow through the exhaust hood (that is, the exhaust volume rate) is the most important factor for all types of hoods. For local, side, downdraft, and canopy hoods, the location is equally important because the rate of airflow is based on the relative distance between the hood and the source. The shape of the exhaust hood is another design consideration. If the hood shape is not selected properly, considerable static pressure losses may result.

Rate of Airflow — Two approaches used in minerals processing operations to determine the rate of airflow needed through a hood are—

- Air induction
- Control velocity

Air Induction — The air-induction concept is based on the theory that when granular material falls through the air each solid particle imparts some momentum to the surrounding air. Due to this energy transfer, a stream of air travels with the material. Unless the air is removed it will escape through all openings upon material impact, carrying the fine dust particles with it. For adequate control of dust emissions, the exhaust air volume rate must be equal to or greater than the air-induction rate.

The air-induction phenomenon is of great significance in calculating exhaust volumes. The concept can be applied to many transfer points normally found in minerals processing operations because the calculations are based on variables such as the material feed rate, its height of free fall, its size, and its bulk density.

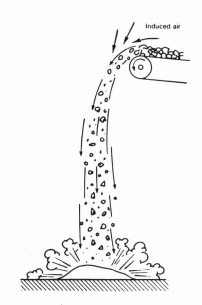

Air-Induction Approach

The exhaust volumes are calculated before the exhaust hood is designed and placed. An approach suggested by Anderson is the most commonly used in the industry today. It is based on the results of a comprehensive laboratory study made by Dennis at Harvard School of Public Health. In its simplified form, it is—

$$Q_{ind} = 10 \times A_u \times \sqrt[3]{\frac{RS^2}{D}}$$

where:

Q_{ind} =   volume of induced air, ft^3/min

A_u =   enclosure open area at upstream end (point where air is induced into the system by action of the falling material), ft^2

R =   rate of material flow rate, ton/h

S =   height of free fall of material, ft

D =   average material diameter, ft

Due to the approximate nature of the formula, Anderson recommends that—

$$Q_{ind} = Q_E$$

where:

Q_E =   required exhaust volume, ft^3/min

The most important parameter in the equation is A_u—the opening through which the air induction occurs. The tighter the enclosure, the smaller the value of A_u and, hence, the smaller the exhaust volume.

Although Anderson's approach can be widely applied, it may not be appropriate for some special operations or situations. When Anderson's approach cannot be applied, the control-velocity approach should be used.

Control Velocity — In the control-velocity approach, the exhaust hood is designed before exhaust volumes are computed. This approach is based on the principle that, by creating sufficient airflow past a dust source, the dusty air can be directed into an exhaust hood. The air velocity required to overcome the opposing air

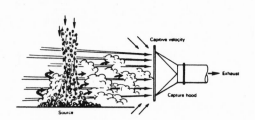

Control-Velocity Approach

currents and capture the dusty air is known as capture velocity.

Dallavalle investigated the air-velocity pattern in a space adjoining an exhaust/suction opening and developed the following equation to determine exhaust volume:

$$Q = V_X (10X^2 + A)$$

where:

Q = exhaust volume, ft^3/min

V_X = centerline velocity (i.e., capture velocity) at distance X from hood, ft/min

X = distance outwards along the hood axis, ft

A = area of hood face opening, ft^2

Capture velocities for some typical operations are provided in the table on the following page.

Location of the Exhaust Hood — The location of the exhaust hood is important in achieving maximum dust-capture efficiency at minimum exhaust volumes. When the control-velocity approach is used, the location of the hood is critical because exhaust volume varies in relation to the location and size of the exhaust hood. The location of the exhaust hood is not as critical when the air-induction approach is used.

The air-induction approach requires the hood to be located as far from the material impact point as possible to—

● Prevent capturing coarse dust particles, which settle quickly

● Capture only fine, predominantly respirable dust

● Reduce unnecessary transport of coarse dust through ductwork and thus reduce dust settling in horizontal duct runs

● Reduce dust loading (dust concentration) in the exhaust gases

● Minimize subsequent cleaning and disposal of the collected dust

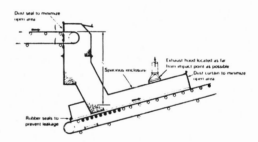

Location of Hood Using Air-Induction Approach

• Prevent capture of valuable products, especially in ore-concentrating operations

The control-velocity approach requires the hood to be located as close to the source as possible to—

• Maximize the hood capture efficiency for a given volume

• Reduce the exhaust volume requirements

• Enclose the source as much as possible

Shape of the Exhaust Hood — Sizable pressure losses may occur if the shape of the exhaust hood is not designed properly. These pressure losses are due to the mutual conversion of static and velocity pressures.

The following points should be considered in selecting the shape of the hood:

• The exhaust hood shape with the highest coefficient of entry value, C_e, or the lowest hood entry loss factor, h_e, should be selected. Various values of C_e and h_e are described in the following table:

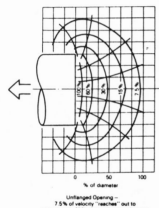

Unflanged Opening –
7.5% of velocity "reaches" out to
approximately 95% of duct diameter distance.

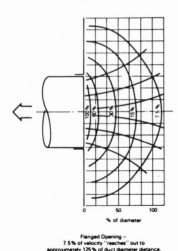

Flanged Opening –
7.5% of velocity "reaches" out to
approximately 125% of duct diameter distance.

Effect of Flanged Opening

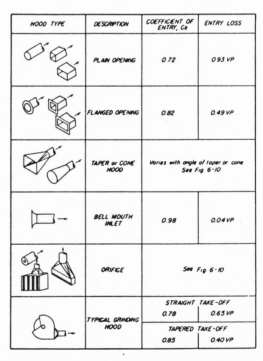

HOOD TYPE	DESCRIPTION	COEFFICIENT OF ENTRY, Ce	ENTRY LOSS
	PLAIN OPENING	0.72	0.93 VP
	FLANGED OPENING	0.82	0.49 VP
	TAPER or CONE HOOD	Varies with angle of taper or cone See Fig 6-10	
	BELL MOUTH INLET	0.98	0.04 VP
	ORIFICE	See Fig 6-10	
	TYPICAL GRINDING HOOD	STRAIGHT TAKE-OFF 0.78	0.65 VP
		TAPERED TAKE-OFF 0.85	0.40 VP

Reprinted by permission from the Committee on Industrial Ventilation, Lansing, MI, 18th Edition.

● Wherever possible, the hood should be flanged to eliminate airflow from zones containing no contaminants. This measure can reduce exhaust air volume up to 25%. For most applications, the flange width should not exceed 6 in.

Ductwork

The ductwork transports the dust captured by the exhaust hood to a dust collector. Efficient transport of captured dust is necessary for effective and reliable system operation.

Ductwork Design

Ductwork design includes the selection of duct sizes based on the velocity necessary to carry the dust to the collector without settling in the duct. From this information, pressure losses in the duct and exhaust air volumes can be calculated and used to determine the size and type of fan, as well as the speed and size of motor.

Before detailed design of the ductwork is begun, the following information should be available:

● A process flowsheet of the operation indicating—

   - Type, size, and speed of the bulk material handling or processing equipment used

● A line diagram of the dust collection system indicating—

   - Exhaust hood and exhaust volumes required for each piece of equipment, each transfer point, and each duct network

   - Each branch and section of the main duct, identified either by number or letter

● A general layout of the facility showing—

   - All equipment in the plan and elevations

   - The ductwork route and location of the exhaust hood

   - Location of the dust collector and the fan

- A preliminary bill of material containing—

  - Length of each duct

  - Number and type of elbows, transition, and taper pieces, etc.

  - Number and size of "y" branches for each branch and main as identified in the process flowsheet

Proper ductwork design—

- Maintains adequate transport velocities in the duct to prevent particulate settling

- Provides proper air distribution in all branches to maintain designed capture velocities of exhaust hoods

- Minimizes pressure losses, wear, and abrasion of ductwork thus reducing operating costs

Transport Velocities — To prevent dust from settling and blocking the ductwork, transport velocities should range from 3,500 to 4,000 ft/min for most industrial dust (such as granite, silica flour, limestone, coal, asbestos, and clay) and from 4,000 to 5,000 ft/min for heavy or moist dust, such as lead, cement, and quick lime. The table describes minimum transport velocities for different characteristics of dust.

Note: The minimum transport velocity indicated in the table is for guidance only. The design velocity should be estimated by including a safety factor in the above minimum velocities. Estimation of safety factors should consider—

- Material buildup
- Duct damage
- Corrosion of ductwork
- Duct leakage

Material	Minimum Design Velocity (fpm)
Very fine, light dusts	2,000
Fine, dry dusts and powders	3,000
Average industrial dusts	3,500
Coarse dusts	4,000-4,500
Heavy or moist dust loading	4,500 and up

Distribution of Airflow — Proper airflow distribution in each branch is necessary to maintain adequate capture and transport velocities in the system. If air is not properly distributed in a multiple-branch dust collection system, a natural balance will take place. For example, the exhaust volume will be determined by the resistance of the available flow paths, and the branch with the least resistance will carry the most volume. As a result, the desired airflow may not be achieved in each branch.

The system should be balanced to ensure desired airflow distribution. In other words, all branches entering a junction must have equal static pressures at the designed flow. Two methods available to balance the system are—

- Air balance without blast gates
- Air balance with blast gates

Air Balance Without Blast Gates — This method, often called the static pressure balance method, provides a way to achieve the desired airflow (a balanced system) without the use of dampers or blast gates. Calculation begins at the branch of greatest resistance and proceeds from branch to main, through each section of main, and to the fan. At each junction of two airstreams, the static pressure necessary to achieve desired flow in both streams is matched and, thus, branches are brought into "balance." The static pressures can be balanced at the desired rate of flow by choosing appropriate sizes of ducts, elbow radii, etc.

Air Balance With Blast Gates — This method uses blast gates to achieve the desired airflow at each hood. Calculation begins at the branch of greatest resistance, and pressure drops are calculated through the branch and through the various sections of the main to the fan. No attempt is made to balance the static pressure in the joining airstreams. The joining branches are merely sized to provide the desired transport velocities.

Note: Choosing the branch of greatest resistance is critical in this method. If the choice is incorrect, any branch or branches having a higher resistance will fail to draw the desired volume even when their blast gates are wide open. To prevent this error, all branches that could possibly give the greatest resistance must be checked.

Selection of Balancing Method — Both of the above approaches are common. However, air balance without blast gates normally is selected for processes where highly toxic materials are exhausted so that possible tampering with blast gates will not affect airflow. Air balance with blast gates is selected when exhaust volumes cannot be properly estimated or the system requires some flexibility in varying exhaust volumes.

Note: In the air balance without blast gates method, although calculations are time consuming during the design stage, airflow in the field need not be measured and balanced. In the air balance with blast gates method, the design calculations are fast, but considerable efforts are required in the field to measure and adjust the blast gates to achieve the balance.

Irrespective of the method selected, **additional hoods should not be added** once a multiple hood layout is completed and balanced because they may alter the airflow and make some other hoods totally ineffective. A comparison of both balancing methods is provided in the table on the following page.

Pressure Losses — Pressure losses occur when air travels in a duct. To overcome these pressure losses, power is supplied by the fan and motor. The higher the pressure losses, the greater the motor horsepower requirements.

Pressure losses in a dust collection system occur due to the following:

- Hood entry
- Special duct fittings
- Duct friction
- Air-cleaning devices

Hood-Entry Losses — A loss in pressure occurs when air enters a suction or hood opening. This loss is indicated by the coefficient of entry for the hood (C_e). Several examples of entry coefficients are illustrated.

Losses from Special Duct Fittings — When air travels through the various duct fittings, such as elbows, "y" branches, enlargements, or contractions (tapers), pressure losses occur. Pressure loss across these fittings is expressed in one of two ways:

- As a fraction of the velocity pressure

- In terms of equivalent feet of straight duct (of the same diameter) that will produce the same pressure loss as the fitting

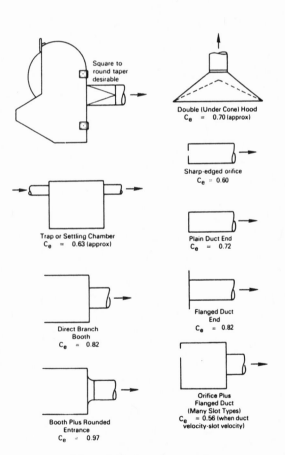

Hood-Entry Losses

Duct-Friction Losses — When air travels in a straight run of duct, pressure losses occur due to the friction between the duct walls and air. Many charts and graphs are available that give friction losses in straight ducts. However, most of them are based on new, clean ducts. The following chart, which allows for a typical amount of roughness, plots four quantities. If any two quantities are known, the other two can be read directly from the chart.

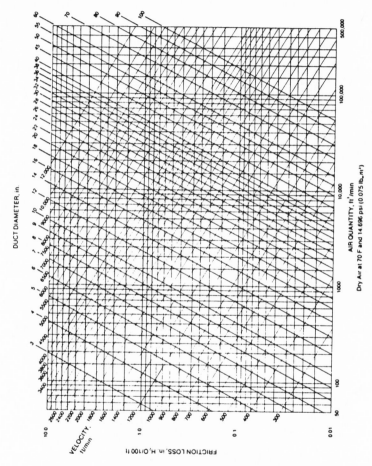

Duct Friction Losses

Reprinted by permission from the Committee on
Industrial Ventilation, Lansing, MI, 18th Edition.

Losses from Air-Cleaning Devices — In addition to pressure losses in the ductwork, the losses in the dust collector must also be known. Although the pressure drop for dust collectors varies widely, data are usually available from manufacturers. More information on dust collectors can be obtained from chapter 4.

Points to Note in Ductwork Design/Layout — To minimize pressure losses, the Industrial Ventilation Manual recommends the following guidelines for ductwork design:

- All branches should enter the main at a 30° angle; wherever possible, the velocity should match that of the incoming gas stream.

- Duct size changes should be kept to a minimum. If needed, they should be gradual.

- Wherever possible, a circular duct should be used instead of a rectangular duct to maintain uniform velocity distribution and prevent settling of material in the ductwork.

- Wherever possible, flanges should be provided to minimize hood entry losses.

- The centerline radii of all elbows should be at least twice the diameter of the duct.

The illustrations on the following page provide examples of ductwork design.

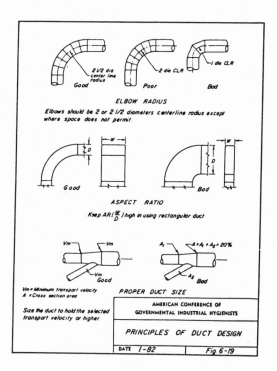

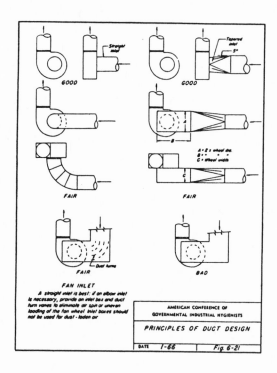

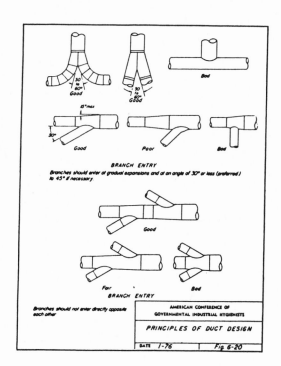

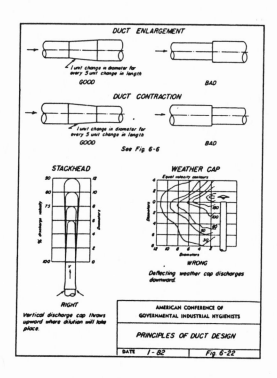

Reprinted by permission from the Committee on
Industrial Ventilation, Lansing, MI, 18th Edition.

Example Problem

This example, which illustrates the balancing of ductwork, is provided to aid understanding of the detailed ductwork design procedure. In the example, the balancing of ductwork is based on the air balance without blast gate method, and the resistances are based on the equivalent foot basis.   This approach is one of several available for balancing ductwork; however, an understanding of this approach should facilitate understanding of other approaches. Information on other approaches can be obtained from the sources provided in the references.

The Problem

Design a dust collection system for an industrial sand-handling facility.

Information Provided

- Process flowsheet and schematic of the dust collection system

- Minimum transport velocity = 3,500 fpm

- Necessary exhaust volumes

- Description and materials

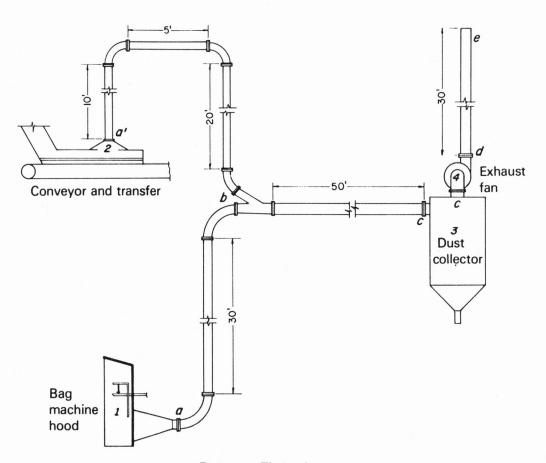

Process Flowsheet

Details of Operation

Description	Number	Minimum Exhaust Volume (cfm)
Bag Machine Hood	1	800
Conveyor Transfer Point	2	300
Bag House Dust Collector	3	-
Fan	4	

Description of Material

No. of Branch or Main	Airflow Required (cfm)	Straight Run (ft)	Number of Elbows	Number of Entries
1-b	800	30	2-90°	--
2-b	300	35	3-90°	--
b-c	1,100	50	--	1
c-d	1,100	0	1-90°	--
d-e	1,100	30	--	--

After gathering the above information you can start to fill in the worksheet table:

Column	Entry	Explanation
1	1-b	Section of system to be worked on from bag hood to Y junction
2	6.5 in.	Based on minimum transport velocity and minimum air volume required using Q = VA formula A = 0.27 ft$_2$, this gives a duct size of 6.5 in.
3	0.27	Duct area
4	800	Air volume required, as calculated from Andersen, or others, or determined by past experience or testing
6	3500	Air velocity determined by Q/A = V
7	30	Straight runs measured from prints or on site
8	2	Number of elbows determined from schematic installation
9	14	Taken from table ____ for an elbow with a centerline radius of 2.0 times the duct diameter. This number is multiplied times the number of elbows.
10	44	Sum of straight run length (col. 7) and equivalent length (col. 9)
11	2.7	Frictional losses read directly from Figure ____, look at duct diameter vs. duct velocity and read frictional loss per 100 ft. of straight total duct length
12	1.19	Total duct length (col. 10) x frictional loss per 100 ft. of duct (col. 11) divided by 100
13	0.77	Velocity pressure of air in duct, $VP = (V/P)^2$
14	0.5	Entry loss for hood (given)
15	1.5	The total exhaust hood loss. This represents the amount of energy required to get the air to flow into the hood (1.0 VP) plus the specific hood entry loss (col. 14).
16	1.16	Product of cols. 13 and 15
17	1.16	The governing static pressure in branch 1-b (col. 12 + col. 16).

Repeat this procedure for each branch circuit, when you reach a junction of two branch circuits the balanced pressure method requires that the governing static pressure at the junction be within 5% of one another. If this is not the case, as in our example, then design parameters must be altered to achieve balance. Several things might be done:

- lower the resistance in branch 2-b by increasing duct size, or reducing air volume
- increase the resistance in branch 1-b by decreasing duct size or increasing air volume

Engineering and economic judgment should be used to make this decision; for instance, whether you can use additional air volume or your fan cannot handle the static pressure may dictate the way in which you choose to balance your system. In the example we chose to increase the resistance in branch 1-b to achieve balance. This step may require several trial and error attempts until you become familiar with the process.

Continue filling in the work sheet using the governing static pressure column to keep a running total of pressure (note that pressures in series circuits are additive; in parallel circuits they are not). The governing pressure in that branch is used.

Finally, as the air exits the exhaust stack of the fan, the velocity pressure of the air is converted back to static pressure and results in the recovery of that energy which is subtracted from the governing static pressure of the system.

WORKSHEET
Example Problem

1	2	3	4	5	6	7	8		9	10	11	12	13	14	15	16	17	18	19
					From Sect. 5				From Fig. 4	Col 7 + Col 9	From Fig. 3	Col 10 x Col 11 100	From Col 6 & Fig. 2	From Fig. 1	100 + Col 14	Col 13 x Col 15	Col 12 + Col 16	At junction	
No. of br. or main	Dia. duct in in.	Area duct in sq. ft.	Air Volume cfm			Length of Duct in Feet													
			In branch	In main	Vel. in fpm	Straight runs	Number of Elbows	Entries	Equiv. length	Total length	Per 100	Of run	One VP	Entry loss (VP)	Hood suct. (VP)	Hood suct.	Static press.	Gov. SP	Corr. cfm
1-b	6.5	.27	800	-	3500	30	2	-	14	44	2.7	1.19	.77	.5	1.5	1.16	2.35		
2-b	4	.08	300	-	3750	35	3		12	47	4.25	2.0	.88	.4	1.4	1.23	3.23	3.23	

Note: Column 17, resistance of branch 1-b is less than 2-b; this will cause more air than required at the bag machine hood, and less than required at the conveyor transfer point. To balance the flows, recalculate branch 1-b using smaller duct diameter. (Or, alternatively, reduce resistance in 2-b provided duct velocities do not drop below design criteria.)

1	2	3	4	5	6	7	8		9	10	11	12	13	14	15	16	17	18	19
1-b	6.0	.196	800	-	4081	30	2	-	14	44	4.2	1.85	1.01	.5	1.5	1.52	-3.37	3.37	800

Branches 1-b and 2-b are now within +5% and are considered balanced.

1	2	3	4	5	6	7	8		9	10	11	12	13	14	15	16	17	18	19
b-c	7.5	.31	1100	1100	3548	50	-	1	7	57	2.3	1.31	.78	-	-	-	-4.68		
3 Collector resistance from manufacture												2.0					-6.68		
c-d	7.5	.31	1100	1100	3548	0	1	0	10	10	2.3	.23	.78	-	-	-	-6.91		
d-e	7.5	.31	1100	1100	3548	30	0	0	0	30	2.3	.69	.78	-	-	-	+7.60		

Fan SP = sum of resistances to fan – VP (VP converted to SP)
= 7.60 –.78
= 6.82" WG @ 1100 cfm

Wet Dust Suppression Systems

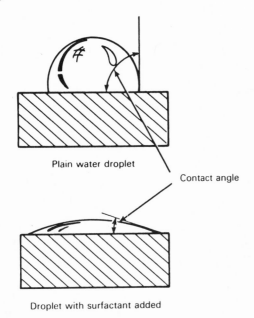

Plain water droplet

Contact angle

Droplet with surfactant added

Contact Angle of a Water Droplet

Wet dust suppression systems wet the entire product stream so that it generates less dust. This also prevents dust from becoming airborne. Effective wetting of the material can be achieved by—

- **Static Spreading** — The material is wetted while stationary. The diameter and contact angle of water droplets are important factors in static spreading.

  The surface coverage can be increased by reducing either the droplet diameter or its contact angle.

- **Dynamic Spreading** — The material is wetted while moving. The surface tension of the liquid, the droplet diameter, the material size, and the droplet impact velocity are important variables in dynamic spreading.

  The surface coverage can be increased either by reducing the surface tension or by increasing the impact velocity.

One of these two water spreading methods can be emphasized at the expense of the other, depending on the needs of the system. For example, both static and dynamic spreading of a droplet can be increased by reducing the surface tension and thus decreasing the droplet diameter. However, the impact velocity of smaller droplets decreases faster due to frictional drag and less momentum, which, in turn, reduces dynamic spreading. An optimum droplet diameter for maximum material surface coverage must therefore be determined.

Factors Affecting Surface Wetting

Droplet Size

Surface wetting can be increased by reducing the droplet diameter and increasing the number of droplets. This can be achieved by reducing the surface tension/contact angle. The surface tension of pure water is 72.6 dyne/cm. It can be reduced from 72.6 to 28 dyne/cm by adding minute quantities of surfactants. This reduction in surface tension (or contact angle) results in—

- Reduced droplet diameter
- An increase in the number of droplets
- A decrease in the contact angle

Impact Velocity

Surface wetting can be increased by increasing the impact velocity. **Impact velocity can be increased by increasing the system's operating pressure.**

Note: A droplet normally travels through turbulent air before it impacts on the material surface. Due to the frictional drag of the turbulent air, the impact velocity of the droplet is less than its discharge velocity from the nozzle. Moreover, small droplets lose velocity faster than large ones. To cover the greatest surface area, the best impact velocity for a given droplet diameter must be determined for each operation.

Wet suppression systems fall into three categories:

- **Plain Water Sprays** — This method uses plain water to wet the material. However, it is difficult to wet most surfaces with plain water due to its high surface tension.

- **Water Sprays with Surfactant** — This method uses surfactants to lower the surface tension of water. The droplets spread further and penetrate deeper into the material pile.

- **Foam** — Water and a special blend of surfactant make the foam. The foam increases the surface area per unit volume, which increases wetting efficiency.

Types of Wet Dust Suppression Systems

Advantages and Disadvantages

Advantages	Disadvantages

Plain Water Sprays

- It is probably the least expensive method of dust control.

- The system is simple to design and operate.

- A limited carryover effect at subsequent transfer points is possible.

- When good mixing of water and material can be achieved, dust generation can be reduced effectively.

- Enclosure tightness is not essential.

- Water sprays cannot be used for products that cannot tolerate excessive moisture.

- Water sprays cannot be used when temperatures fall below freezing.

- Usually, dust control efficiency is low, unless large quantities of water are used.

- Freeze protection of all hardware is necessary.

- Careful application at transfer points that precede a screen is required to prevent blinding.

Water Sprays With Surfactants

- This method is used when surfactants are tolerated but excessive moisture is not acceptable.

- In some cases, dust control efficiency is higher than with plain water sprays.

- Equivalent efficiency is possible with less water.

- Capital and operating costs are higher than water-spray systems.

- Careful application at transfer points that precede a screen is required to prevent blinding.

- Equipment such as the pump and the proportioning equipment used to meter the flow of surfactant require maintenance.

- Freeze protection of all hardware is necessary.

Foam

- When good mixing of foam and product stream can be achieved, dust control efficiency is greater than water with surfactants.

- Moisture addition is usually less than 0.1% of the material weight.

- Operating costs are higher than with finely atomized water-spray systems.

- The product is contaminated with surfactants.

- Careful application at transfer points that precede a screen is required to prevent blinding.

In this approach, very fine water droplets are sprayed into the dust after it is airborne. When the water droplets and dust particles collide, agglomerates are formed. When these agglomerates become too heavy to remain airborne, they settle.

Airborne Dust Capture Systems

Collision Between Dust Particle and Water Droplet

Coalescence or Adhesion Between Dust Particle and Water Droplet

The collision between dust particles and water droplets occurs due to the following three factors:

- Impaction/interception
- Droplet size/particle size
- Electrostatic forces

Impaction/Interception

When a dust particle approaches a water droplet, the airflow may sweep the particle around the droplet or, depending on its size, trajectory, and velocity, the dust particle may strike the droplet directly, or barely graze the droplet, forming an aggregate.

Droplet Size/Particle Size

Droplets and particles that are similar in size have the best chance of colliding. Droplets smaller than dust particles or vice versa may never collide but just be swept around one another.

Electrostatic Forces

The presence of an electrical charge on a droplet affects the path of a particle around the droplet. When particles have an opposite or neutral charge, collision efficiency is increased.

Factors Affecting Collision

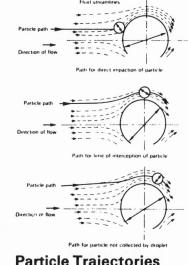

Particle Trajectories Around a Water Droplet

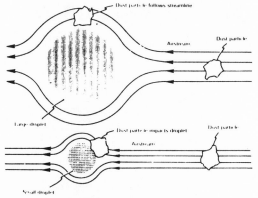

Effect of Droplet Size
Schowengerdt and Brown

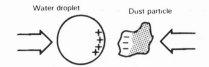

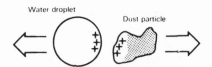

**Oppositely Charged Droplet
and Particle Attract
Each Other**

**Similarly Charged Droplet
and Particle Oppose
Each Other**

Types of Airborne Dust Capture Systems

Airborne dust capture systems can be simple or quite complex. Basically, they fall into two broad groups:

- Finely atomized water sprays
- Electrostatically charged fogs

Finely Atomized Water Sprays

Finely atomized water sprays are normally used at transfer points without excessive turbulence or when the velocity of dust dispersion is less than 200 ft/min. The optimum droplet size, water usage, relative velocity, and number and location of nozzles depend on the conditions at individual transfer points.

Electrostatically Charged Fogs

Electrostatically charged fog uses charged water droplets to attract dust particles, which increases collision. The atomized water droplets are charged by induction or direct charging.

Design of a Water-Spray System

The spray nozzle is the heart of a water-spray system. Therefore, the physical characteristics of the spray are critical. Factors such as droplet size distribution and velocity, spray pattern and angle, and water flow rate and pressure all vary depending on the nozzle selected. Following is a general discussion of these important factors:

- **Droplet Size** — The nozzle's droplet size distribution is the most important variable for proper dust control. The droplet size decreases as the operating pressure increases. Information about

Advantages and Disadvantages

Advantages	Disadvantages

Finely Atomized Water Sprays

- Water requirements are low—typically 5 to 20 gal/h per nozzle.

- Moisture addition to the product is quite low—typically less than 0.1% of the material weight.

- The material is not chemically contaminated.

- The system can be economical.

- Tight enclosures are needed for effective system operation.

- The system may not be effective either in highly turbulent environments or when the dust dispersion rate is more than 200 ft/min.

- Requires good droplet to particle size match for effective control.

Electrostatically Charged Fogs

- Electrostatic fogs can be effective if the dust cloud carries predominantly positive or negative charges.

- The material does not become chemically contaminated.

- Moisture addition to the product is generally less than 0.5% by weight.

- These systems are not recommended for underground coal mines or other gassy applications where explosions can be triggered by sparks.

- Capital costs are high.

- These systems require high-voltage equipment.

- Maintenance of electrical insulation is critical for safe working conditions.

the droplet size data at various operating pressures can be obtained from the nozzle manufacturer. For wet dust suppression systems, coarse droplets (200-500 μm) are recommended. For airborne dust capture systems, very fine droplets (10-150 μm) may be required. The fine droplets usually are generated by fogging nozzles, which may use either compressed air or high-pressure water to atomize water in the desired droplet range.

- **Droplet Velocity** — Normally, higher droplet velocities are desirable for both types of dust control through water sprays. Information on the droplet velocity can be obtained from the nozzle manufacturer.

- **Spray Pattern** — Nozzles are categorized by the spray patterns they produce:

  - Solid-cone nozzles produce droplets that maintain a high velocity over a distance. They are useful for providing a high-velocity spray when the nozzle is located distant from the area where dust control is desired.

  - Hollow-cone nozzles produce a spray pattern in the form of a circular ring. Droplet range is normally smaller than the other types of nozzles. They are useful for operations where dust is widely dispersed.

  - Flat-spray nozzles produce relatively large droplets that are delivered at a high pressure. These nozzles are normally useful for wet dust suppression systems (i.e., preventive type systems).

  - Fogging nozzles produce a very fine mist (a droplet size distribution ranging from submicron to micron). They are useful for airborne dust control systems.

- **Spray Angle** — Each nozzle has a jet spray angle. The size of this angle is normally available from the manufacturer. A knowledge of spray angle and spray pattern is essential to determine the area of coverage and, therefore, the total number of nozzles needed.

Solid-Cone Nozzle

Hollow-Cone Nozzle

Flat-Spray Nozzle

Fogging Nozzle

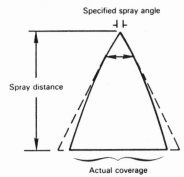

Spray Angle

From Bureau of Mines open File Report 145-82, Guide Book for Dust Control in Underground Mining, December 1981.

- **Flow Rate** — The flow rate of water through a nozzle depends on the operating pressure. The flow rate and operating pressure are related as follows:

$$\text{Water flow rate} = K\sqrt[2]{\text{operating pressure}}$$

  where K = nozzle constant

  A knowledge of the water flow rate through the nozzle is necessary to determine the percentage of moisture added to the material stream.

The following factors should be considered in selecting the nozzle location:

- It should be readily accessible for maintenance.

- It should not be in the path of flying material.

- For wet dust suppression systems, nozzles should be **upstream** of the transfer point where dust emissions are being created. Care should be taken to locate nozzles for best mixing of material and water. For airborne dust capture, nozzles should be located to provide **maximum time** for the water droplets to interact with the airborne dust.

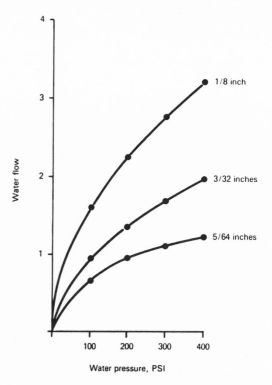

Water Flow/Pressure

Water Flow and Compressed Airflow Rates

Once the nozzle is selected, its spray pattern and area of coverage can be used to determine water flow rate and/or compressed airflow rates and pressure requirements. This information is normally published by the nozzle manufacturer. These must be carefully coordinated with the maximum allowable water usage. Water flow rates will be highly variable depending on the size and type of material, the type of machinery, and the throughput of material.

Piping Design

The piping should be designed so that each nozzle receives water or compressed air at specified flow rates and pressures. Drains must be provided at the lowest point in each subcircuit of the piping system to flush the air and water lines in winter months. Heat tapes and insulation must also be provided at locations where the temperature may drop below 32° F. The heat tracing tape should be able to provide approximately 4 watts per linear foot for water pipes up to 2 in. in diameter. The pump and other hardware, such as valves and gauges, should also be heat traced and insulated to prevent freezing during winter months.

Instruments

Pressure and flow gauges are recommended to monitor system performance. These instruments should be located as close to the point of application as possible. Liquid-filled pressure gauges and rotameter-type flowmeters are satisfactory and quite inexpensive.

For situations where it is desirable to activate wet suppression systems only when the material is flowing (for example, if the belt conveyor is running empty, water sprays need not be on), a solenoid-activated valve may be installed in the water line. The solenoid can be activated by instruments such as the level controller or flow sensor. This measure will reduce water usage, reduce maintenance and cleanup, and reduce or prevent freezeup problems.

Pump and Compressor Selection

An appropriate pump and compressor (where applicable) should be selected once the airflow and water flow rates and pressure are determined.

An approximate method of determining the proper pumping energy for water at 40:1 efficiency is—

$$\text{Pump HP} = \frac{1.40}{10,000} \times p \times q$$

where:

p   =   pressure drop in water lines, psig

q   =   water flow rate, gal/min

An approximate method of selecting a compressor is by assuming that—

One horsepower of compressor can provide approximately 4 std ft^3/min of compressed air, at 100 psig pressure.

Chapter 4

Collecting and Disposing of Dust

What Is a Dust Collector?

After dust-filled air has been captured by a dry dust collection system, it must be separated, collected, and disposed of. The dust collector separates dust particles from the airstream and discharges cleaned air either into the atmosphere or back into the workplace.

Necessity for Dust Collectors

Cleaning dust from the air is necessary to—

- Reduce employee exposure to dust

- Comply with health and air emission standards

- Reduce nuisance and dust exposure to neighbors

- Recover valuable products from the air

Types of Dust Collectors

Five principal types of industrial dust collectors are—

- Inertial separators
- Fabric collectors
- Wet scrubbers
- Electrostatic precipitators
- Unit collectors

Types of Inertial Separators

Inertial separators separate dust from gas streams using a combination of forces, such as centrifugal, gravitational, and inertial. These forces move the dust to an area where the forces exerted by the gas stream are minimal. The separated dust is moved by gravity into a hopper, where it is temporarily stored.

The three primary types of inertial separators are—

● Settling chambers

● Baffle chambers

● Centrifugal collectors

Neither settling chambers nor baffle chambers are commonly used in the minerals processing industry. However, their principles of operation are often incorporated into the design of more efficient dust collectors.

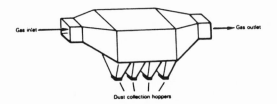

Settling Chamber

Settling Chambers

A settling chamber consists of a large box installed in the ductwork. The sudden expansion of size at the chamber reduces the speed of the dust-filled airstream and heavier particles settle out.

Settling chambers are simple in design and can be manufactured from almost any material. However, they are seldom used as primary dust collectors because of their large space requirements and low efficiency. A practical use is as precleaners for more efficient collectors.

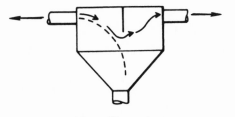

Baffle Chamber

Baffle Chambers

Baffle chambers use a fixed baffle plate that causes the conveying gas stream to make a sudden change of direction. Large-diameter particles do not follow the gas stream but continue into a dead air space and settle. Baffle chambers are used as precleaners for more efficient collectors.

Centrifugal Collectors

Centrifugal collectors use cyclonic action to separate dust particles from the gas stream. In a typical cyclone, the dusty gas stream enters at an angle and is spun rapidly. The centrifugal force created by the circular flow throws the dust particles toward the wall of the cyclone. After striking the wall, these particles fall into a hopper located underneath.

The most common types of centrifugal, or inertial, collectors in use today are—

● Single-cyclone separators

● Multiple-cyclone separators

Single-cyclone separators create a dual vortex to separate coarse from fine dust. The main vortex spirals downward and carries most of the coarser dust particles. The inner vortex, created near the bottom of the cyclone, spirals upward and carries finer dust particles.

Multiple-cyclone separators, also known as multiclones, consist of a number of small-diameter cyclones, operating in parallel and having a common gas inlet and outlet, as shown in the figure. Multiclones operate on the same principle as cyclones— creating a main downward vortex and an ascending inner vortex.

Multiclones are more efficient than single cyclones because they are longer and smaller in diameter. The longer length provides longer residence time while the smaller diameter creates greater centrifugal force. These two factors result in better separation of dust particulates. The pressure drop of multiclone collectors is higher than that of single-cyclone separators.

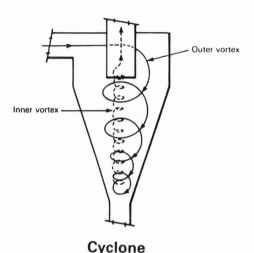

Cyclone

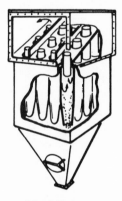

Multiclone

Advantages and Disadvantages — Centrifugal Collectors

Types	Advantages	Disadvantages
Cyclones	• Have no moving parts • Can be used as precleaners to remove coarser particulates and reduce load on more efficient dust collectors • Can be designed to remove a specific size range of particles	• Have low collection efficiency for respirable particulates • Suffer decreased efficiency if gas viscosity or gas density increases • Are susceptible to erosion • Have drastically reduced efficiency due to reduction in airflow rate • Cannot process sticky dust
Multiclones	• Have no moving parts • Are more efficient than single-cyclone separators • Have low pressure drop when used as a precleaner	• Have low collection efficiency for respirable particulates • Are prone to plugging due to smaller diameter tubes • Improper gas distribution may result in dirty gas bypassing several tubes • Cannot process sticky dust • For a given gas volume, occupy more space than single-cyclone separators • Normally have higher pressure drop than single-cyclone separators

Common Operating Problems and Solutions — Cyclones

Symptom	Cause	Solution
● Erosion	● High concentrations of heavy, hard, sharp-edged particles	● Install large-diameter "roughing" cyclone upstream of high-efficiency, small-diameter cyclone. ● Line high-efficiency cyclone with refractory or erosion-resistant material.
● Corrosion	● Moisture and condensation in cyclone	● Keep gas stream temperature above dewpoint. ● Insulate cyclone. ● Use corrosion-resistant material such as stainless steel or nickel alloy.
● Dust buildup	● Gas stream below dewpoint ● Very sticky material	● Maintain gas temperature above dewpoint. ● Install vibrator to dislodge material.
● Reduced efficiency or dirty discharge stack	● Leakage in ductwork of cyclone	● Clean cyclone routinely. ● Check for pluggage and leakage and unplug or seal the ductwork. ● Close all inspection ports and openings.
	● Reduced gas velocity in cyclone	● Check the direction of fan rotation; if rotation is wrong, reverse two of the three leads on motor.

Common Operating Problems and Solutions — Multiclones

Symptom	Cause	Solution
● Erosion	● High concentrations of heavy, hard, sharp-edged particles	● Install cast iron tubes. ● Install a wear shield to protect tubes.
● Overloaded tubes ● Loss of volume in tubes ● Uneven pressure drop across tubes	● Uneven gas flow and dust distribution	● Install turning vanes in elbow, if elbow precedes inlet vane.
● Plugging in inlet vanes, clean gas outlet tubes, and discharge hopper	● Low gas velocity ● Uneven flow distribution ● Moisture condensation ● Overfilling in discharge hopper	● Install turning vanes in elbow inlet. ● Insulate multiclone. ● Install bin-level indicator in collection hopper. ● Empty hopper more frequently.
● Reduced efficiency or dirty gas stack	● Leakage in ductwork ● Leakage in multiclone	● Seal all sections of ductwork and multiclone to prevent leaks.

Startup/Shutdown Procedures — Centrifugal Collectors

Type	Startup	Shutdown
Cyclones	1. Check fan rotation. 2. Close inspection doors, connections, and cyclone discharge. 3. Turn on fan. 4. Check fan motor current. 5. Check pressure drop across cyclone.	1. Allow exhaust fan to operate for a few minutes after process shutdown until cyclone is empty. 2. If combustion process is used, allow hot, dry air to pass through cyclone for a few minutes after process shutdown to avoid condensation. 3. Turn off exhaust fan. 4. Clean discharge hopper.
Multiclones	1. Conduct same startup procedures as cyclones. 2. At least once a month, measure airflow by conducting a Pitot traverse across inlet to determine quantity and distribution of airflow. 3. Record pressure drop across multiclone. 4. If flow is significantly less than desired, block off rows of cyclone to maintain the necessary flow per cyclone.	1. Conduct same shutdown procedures as cyclones.

Preventive Maintenance Procedures — Centrifugal Collectors

Type	Frequency	Procedure
Cyclones	Daily	• Record cyclone pressure drops. • Check stack (if cyclone is only collector). • Record fan motor amperage. • Inspect dust discharge hopper to assure dust is removed.
	Weekly	• Check fan bearings. • Check gaskets, valves, and other openings for leakage.
	Monthly	• Check cyclone interior for erosion, wear, corrosion, and other visible signs of deterioration.
Multiclones	Daily	• Same as cyclones.
	Weekly	• Same as cyclones.
	Monthly	• Check multiclone interior for erosion, wear, corrosion, and improper gas and dust distribution. • Inspect individual cyclones and ducts for cracks caused by thermal expansion or normal wear.

Commonly known as baghouses, fabric collectors use filtration to separate dust particulates from dusty gases. They are one of the most efficient and cost-effective types of dust collectors available and can achieve a collection efficiency of more than 99% for very fine particulates.

Fabric Collectors

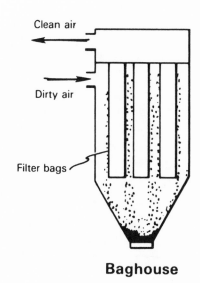

Baghouse

Dust-laden gases enter the baghouse and pass through fabric bags that act as filters. The bags can be of woven or felted cotton, synthetic, or glass-fiber material in either a tube or envelope shape.

How They Work

The high efficiency of these collectors is due to the dust cake formed on the surfaces of the bags. The fabric primarily provides a surface on which dust particulates collect through the following four mechanisms:

- **Inertial Collection** — Dust particles strike the fibers placed perpendicular to the gas-flow direction instead of changing direction with the gas stream.

- **Interception** — Particles that do not cross the fluid streamlines come in contact with fibers because of the fiber size.

- **Brownian Movement** — Submicron particles are diffused, increasing the probability of contact between the particles and collecting surfaces.

- **Electrostatic Forces** — The presence of an electrostatic charge on the particles and the filter can increase dust capture.

A combination of these mechanisms results in formation of the dust cake on the filter, which eventually increases the resistance to gas flow. The filter must be cleaned periodically.

Types of Baghouses

As classified by cleaning method, three common typesof baghouses are—

- Mechanical shaker
- Reverse air
- Reverse jet

Mechanical Shaker

In mechanical-shaker baghouses, tubular filter bags are fastened onto a cell plate at the bottom of the baghouse and suspended from horizontal beams at the top. Dirty gas enters the bottom of the baghouse and passes through the filter, and the dust collects on the inside surface of the bags.

Cleaning a mechanical-shaker baghouse is accomplished by shaking the top horizontal bar from which the bags are suspended. Vibration produced by a motor-driven shaft and cam creates waves in the bags to shake off the dust cake.

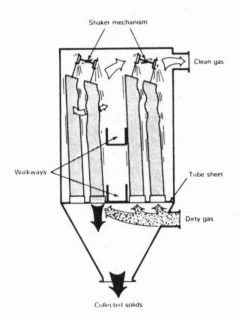

Mechanical-Shaker Baghouse

Shaker baghouses range in size from small, hand-shaker devices to large, compartmentalized units. They can operate intermittently or continuously. Intermittent units can be used when processes operate on a batch basis—when a batch is completed, the baghouse can be cleaned. Continuous processes use compartmentalized baghouses; when one compartment is being cleaned, the airflow can be diverted to other compartments.

In shaker baghouses, there must be no positive pressure inside the bags during the shake cycle. Pressures as low as 0.02 in. wg can interfere with cleaning.

Air-to-Cloth Ratio:
The volume of gas flow passed per unit area of the bag.

The air-to-cloth ratio for shaker baghouses is relatively low, hence the space requirements are quite high. However, because of the simplicity of design, they are popular in the minerals processing industry.

Reverse Air

In reverse-air baghouses, the bags are fastened onto a cell plate at the bottom of the baghouse and suspended from an adjustable hanger frame at the top. Dirty gas flow normally enters the baghouse and passes through the bag from the inside, and the dust collects on the inside of the bags.

Reverse-air baghouses are compartmentalized to allow continuous operation. Before a cleaning cycle begins, filtration is stopped in the compartment to be cleaned. Bags are cleaned by injecting clean air into the dust collector in a reverse direction, which pressurizes the compartment. The pressure makes the bags collapse partially, causing the dust cake to crack and fall into the hopper below. At the end of the cleaning cycle, reverse airflow is discontinued, and the compartment is returned to the main stream.

The flow of the dirty gas helps maintain the shape of the bag. However, to prevent total collapse and fabric chafing during the cleaning cycle, rigid rings are sewn into the bags at intervals.

Space requirements for a reverse-air baghouse are comparable to those of a shaker baghouse; however, maintenance needs are somewhat greater.

Reverse Jet

In reverse-jet baghouses, individual bags are supported by a metal cage, which is fastened onto a cell plate at the top of the baghouse. Dirty gas enters from the bottom of the baghouse and flows from outside to inside the bags. The metal cage prevents collapse of the bag.

Bags are cleaned by a short burst of compressed air injected through a common manifold over a row of bags. The compressed air is accelerated by a venturi nozzle mounted at the top of the bag. Since the duration of the compressed-air burst is short (0.1 s), it acts as a rapidly moving air bubble, traveling through the entire length of the bag and causing the bag surfaces to flex. This flexing of the bags breaks the dust cake, and the dislodged dust falls into a storage hopper below.

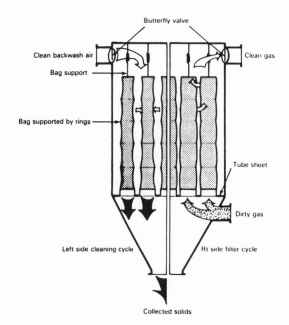

Reverse-Air Baghouse

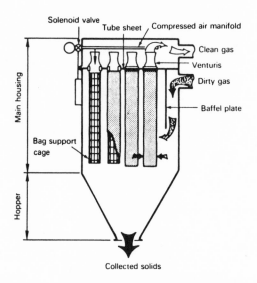

Reverse-Jet Baghouse

Reverse-jet dust collectors can be operated continuously and cleaned without interruption of flow because the burst of compressed air is very small compared with the total volume of dusty air through the collector. Because of this continuous-cleaning feature, reverse-jet dust collectors are usually not compartmentalized.

The short cleaning cycle of reverse-jet collectors reduces recirculation and redeposit of dust. These collectors provide more complete cleaning and reconditioning of bags than shaker or reverse-air cleaning methods. Also, the continuous-cleaning feature allows them to operate at higher air-to-cloth ratios, so the space requirements are lower.

Cartridge Collectors

Cartridge collectors are another commonly used type of dust collector. Unlike baghouse collectors, in which the filtering media is woven or felt bags, this type of collector employs perforated metal cartridges that contain a pleated, nonwoven filtering media. Due to its pleated design, the total filtering surface area is greater than in a conventional bag of the same diameter, resulting in reduced air to media ratio, pressure drop, and overall collector size.

Cartridge collectors are available in single use or continuous duty designs. In single-use collectors, the dirty cartridges are changed while the collector is off. In the continuous duty design, the cartridges are cleaned by the conventional pulse-jet cleaning system.

Advantages and Disadvantages — Baghouses

Types	Advantages	Disadvantages
Mechanical-shaker baghouses	• Have high collection efficiency for respirable dust • Can use strong woven bags, which can withstand intensified cleaning cycle to reduce residual dust buildup • Simple to operate • Have low pressure drop for equivalent collection efficiencies	• Have low air-to-cloth ratio (1.5 to 2 ft/min) • Cannot be used in high temperatures • Require large amounts of space • Need large numbers of filter bags • Consist of many moving parts and require frequent maintenance • Personnel must enter baghouse to replace bags, creating potential for exposure to toxic dust • Can result in reduced cleaning efficiency if even a slight positive pressure exists inside bags
Reverse-air baghouses	• Have high collection efficiency for respirable dust • Are preferred for high temperatures due to gentle cleaning action • Have low pressure drop for equivalent collection efficiencies	• Have low air-to-cloth ratio (1 to 2 ft/min) • Require frequent cleaning because of gentle cleaning action • Have no effective way to remove residual dust buildup • Cleaning air must be filtered • Require personnel to enter baghouse to replace bags, which creates potential for toxic dust exposure
Reverse-jet baghouses	• Have high collection efficiency for respirable dust • Can have high air-to-cloth ratio (6 to 10 ft/min) • Have increased efficiency and minimal residual dust buildup due to aggressive cleaning action • Can clean continuously • Can use strong woven bags • Have lower bag wear • Have small size and fewer bags because of high air-to-cloth ratio • Some designs allow bag changing without entering baghouse • Have low pressure drop for equivalent collection efficiencies	• Require use of dry compressed air • May not be used readily in high temperatures unless special fabrics are used • Cannot be used if high moisture content or humidity levels are present in the exhaust gases

Common Operating Problems and Solutions — Baghouses*

Symptom	Cause	Solution
• High baghouse pressure drop	• Baghouse undersized	• Consult vendor. • Install double bags. • Add more compartments or modules.
	• Bag cleaning mechanism not properly adjusted	• Increase cleaning frequency. • Clean for longer duration. • Clean more vigorously.
	• Shaking not strong enough (MS)	• Increase shaker speed.
	• Compartment isolation damper valves not operating properly (MS, RA)	• Check linkage. • Check valve seals. • Check air supply of pneumatic operators.
	• Compressed air pressure too low (RJ)	• Increase pressure. • Decrease duration and frequency. • Check compressed-air dryer and clean it if necessary. • Check for obstructions in piping.
	• Repressurizing pressure too low (RA)	• Speed up repressurizing fan. • Check for leaks. • Check damper valve seals.
	• Pulsing valves failed (RJ)	• Check diaphragm. • Check pilot valves.
	• Bag tension too tight (RA)	• Loosen bag tension.
	• Bag tension too loose (MS)	• Tighten bags.
	• Cleaning timer failure	• Check to see if timer is indexing to all contacts. • Check output on all terminals.

*MS = mechanical shaker
RA = reverse air
RJ = reverse jet

Symptom	Cause	Solution
	• Not capable of removing dust from bags	• Check for condensation on bags. • Send dust sample and bags to manufacturer for analysis. • Dryclean or replace bags. • Reduce airflow.
	• Excessive reentrainment of dust	• Empty hopper continuously. • Clean rows of bags randomly instead of sequentially (RJ).
	• Incorrect pressure–drop reading	• Clean out pressure taps. • Check hoses for leaks. • Check for proper fluid level in manometer. • Check diaphragm in gauge.
• Dirty discharge at stack	• Bags leaking	• Replace bags. • Tie off leaking bags and replace them later. • Isolate leaking compartment or module.
	• Bag clamps not sealing	• Check and tighten clamps. • Smooth out cloth under clamp and re-clamp.
	• Failure of seals in joints at clean/dirty air connection	• Caulk or weld seams.
	• Insufficient filter cake	• Allow more dust buildup on bags by cleaning less frequently. • Use a precoating on bags (MS, RA).
	• Bags too porous	• Send bag in for permeability test and review with manufacturer.
• High compressed-air consumption (RJ)	• Cleaning cycle too frequent	• Reduce cleaning cycle, if possible.
	• Pulse too long	• Reduce pulsing duration.
	• Pressure too high	• Reduce supply pressure, if possible.
	• Diaphragm valve failure	• Check diaphragm and springs. • Check pilot valve.

Symptom	Cause	Solution
• Reduced compressed-air pressure (RJ)	• Compressed-air consumption too high	• See previous solutions.
	• Restrictions in compressed-air piping	• Check compressed-air piping.
	• Compressed-air dryer plugged	• Replace dessicant in the dryer. • Bypass dryer temporarily, if possible. • Replace dryer.
	• Compressed-air supply line too small	• Consult design.
	• Compressor worn out	• Replace rings. • Check for worn components. • Rebuild compressor or consult manufacturer.
	• Pulsing valves not working	• Check pilot valves, springs, and diaphragms.
	• Timer failed	• Check terminal outputs.
• Moisture in baghouse	• Insufficient preheating	• Run the system with hot air only before process gas flow is introduced.
	• System not purged after shutdown	• Keep fan running for 5 to 10 min after process is shut down.
	• Wall temperature below dewpoint	• Raise gas temperature. • Insulate unit. • Lower dewpoint by keeping moisture out of system.
	• Cold spots through insulation	• Eliminate direct metal line through insulation.
	• Water/moisture in compressed air (RJ)	• Check automatic drains. • Install aftercooler. • Install dryer.
	• Repressurizing air causing condensation (RJ)	• Preheat repressurizing air. • Use process gas as source of repressurizing air.

Symptom	Cause	Solution
• Material bridging in hopper	• Moisture in baghouse	• See previous solutions.
	• Dust stored in hoppers	• Remove dust continuously.
	• Hopper slope insufficient	• Rework or replace hoppers.
	• Screw conveyor opening too small	• Use a wide, flared trough.
• High rate of bag failure, bags wearing out	• Baffle plate worn out	• Replace baffle plate.
	• Too much dust	• Install primary collector.
	• Cleaning cycle too frequent	• Slow down cleaning.
	• Inlet air not properly baffled from bags	• Consult vendor.
	• Shaking too violent (MS)	• Slow down shaking mechanism.
	• Repressurizing pressure too high (RA)	• Reduce pressure.
	• Pulsing pressure too high (RJ)	• Reduce pressure.
	• Cages have barbs (RJ)	• Remove cages and smooth out barbs.

Startup/Shutdown Procedures — Baghouses

Startup	Shutdown
1. For processes generating hot, moist gases, preheat baghouse to prevent moisture condensation, even if baghouse is insulated. (Ensure that all compartments of shaker or reverse-air baghouses are open.)	1. Continue operation of dust-removal conveyor and cleaning of bags for 10 to 20 minutes to ensure good removal of collected dust.
2. Activate baghouse fan and dust-removal conveyor.	
3. Measure baghouse temperature and check that it is high enough to prevent moisture condensation.	

Preventive Maintenance Procedures — Baghouses

Frequency	Procedure
Daily	• Check pressure drop. • Observe stack (visually or with opacity meter). • Walk through system, listening for proper operation. • Check for unusual occurrences in process. • Observe control panel indicators. • Check compressed-air pressure. • Assure that dust is being removed from system.
Weekly	• Inspect screw-conveyor bearings for lubrication. • Check packing glands. • Operate damper valves. • Check compressed-air lines, including line filters and dryers. • Check that valves are opening and closing properly in bag-cleaning sequence. • Spot-check bag tension. • Verify accuracy of temperature-indicating equipment. • Check pressure-drop-indicating equipment for plugged lines.
Monthly	• Check all moving parts in shaker mechanism. • Inspect fans for corrosion and material buildup. • Check drive belts for wear and tension. • Inspect and lubricate appropriate items. • Spot-check for bag leaks. • Check hoses and clamps. • Check accuracy of indicating equipment. • Inspect housing for corrosion.
Quarterly	• Inspect baffle plate for wear. • Inspect bags thoroughly. • Check duct for dust buildup. • Observe damper valves for proper seating. • Check gaskets on doors. • Inspect paint, insulation, etc. • Check screw conveyor for wear or abrasion.
Annually	• Check fan belts. • Check welds. • Inspect hopper for wear.

Wet Scrubbers

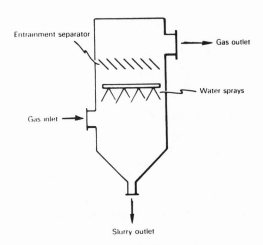

Wet Scrubber

Dust collectors that use liquid are commonly known as wet scrubbers. In these systems, the scrubbing liquid (usually water) comes into contact with a gas stream containing dust particles. The greater the contact of the gas and liquid streams, the higher the dust removal efficiency.

There is a large variety of wet scrubbers; however, all have of three basic operations:

- **Gas-Humidification** — The gas-humidification process conditions fine particles to increase their size so they can be collected more easily.

- **Gas-Liquid Contact** — This is one of the most important factors affecting collection efficiency. The particle and droplet come into contact by four primary mechanisms:

  - Inertial Impaction — When water droplets placed in the path of a dust-laden gas stream, the stream separates and flows around them. Due to inertia, the larger dust particles will continue on in a straight path, hit the droplets, and become encapsulated.

  - Interception — Finer particles moving within a gas stream do not hit droplets directly but brush against them and adhere to them.

  - Diffusion — When liquid droplets are scattered among dust particles, the particles are deposited on the droplet surfaces by Brownian movement, or diffusion. This is the principal mechanism in the collection of sub-micron dust particles.

  - Condensation Nucleation — If a gas passing through a scrubber is cooled below the dew-point, condensation of moisture occurs on the dust particles. This increase in particle size makes collection easier.

- **Gas-Liquid Separation** — Regardless of the contact mechanism used, as much liquid and dust as possible must be removed. Once contact is made, dust particulates and water droplets combine to form agglomerates. As the agglomerates grow larger, they settle into a collector.

The "cleaned" gases are normally passed through a mist eliminator (demister pads) to remove water droplets from the gas stream. The dirty water from the scrubber system is either cleaned and discharged or recycled to the scrubber. Dust is removed from the scrubber in a clarification unit or a drag chain tank. In both systems solid material settles on the bottom of the tank. A drag chain system removes the sludge and deposits it into a dumpster or stockpile.

Wet scrubbers may be categorized by pressure drop (in inches water gauge) as follows:

- Low-energy scrubbers (0.5 to 2.5)

- Low- to medium-energy scrubbers (2.5 to 6)

- Medium- to high-energy scrubbers (6 to 15)

- High-energy scrubbers (greater than 15)

Due to the large number of commercial scrubbers available, it is not possible to describe each individual type here. However, the following sections provide examples of typical scrubbers in each category.

Low-Energy Scrubbers

In the simple, gravity-spray-tower scrubber, liquid droplets formed by liquid atomized in spray nozzles fall through rising exhaust gases. Dirty water is drained at the bottom.

These scrubbers operate at pressure drops of 1 to 2 in. water gauge and are approximately 70% efficient on 10 μm particles. Their efficiency is poor—below 10 μm. However, they are capable of treating relatively high dust concentrations without becoming plugged.

Low- to Medium-Energy Scrubbers

Wet cyclones use centrifugal force to spin the dust particles (similar to a cyclone), and throw the particulates upon the collector's wetted walls. Water introduced from the top to wet the cyclone walls carries these particles away. The wetted walls also prevent dust reentrainment.

Types of Scrubbers

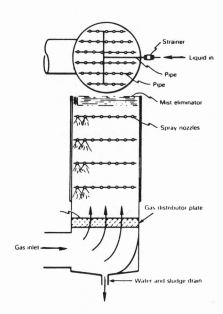

Spray-Tower Scrubber

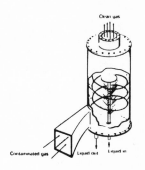

Wet Cyclone

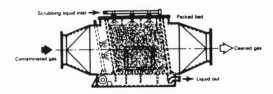

Cross-Flow Scrubber

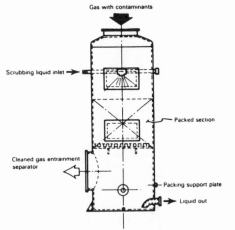

Co-Current-Flow Scrubber

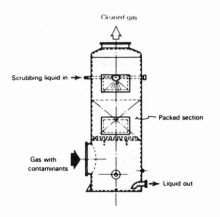

Counter-Current-Flow Scrubber

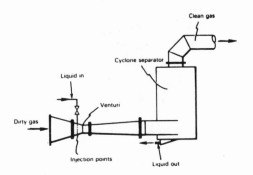

Venturi Scrubber

Pressure drops for these collectors range from 2 to 8 in. water, and the collection efficiency is good for 5 μm particles and above.

Medium- to High-Energy Scrubbers

Packed-bed scrubbers consist of beds of packing elements, such as coke, broken rock, rings, saddles, or other manufactured elements. The packing breaks down the liquid flow into a high-surface-area film so that the dusty gas streams passing through the bed achieve maximum contact with the liquid film and become deposited on the surfaces of the packing elements. These scrubbers have good collection efficiency for respirable dust.

Three types of packed-bed scrubbers are—

- Cross-flow scrubbers
- Co-current flow scrubbers
- Counter-current flow scrubbers

Efficiency can be greatly increased by minimizing target size, i.e., using .003 in. diameter stainless steel wire and increasing gas velocity to more than 1,800 ft/min.

High-Energy Scrubbers

Venturi scrubbers consist of a venturi-shaped inlet and a separator. The dust-laden gases enter through the venturi and are accelerated to speeds between 12,000 and 36,000 ft/min. These high-gas velocities immediately atomize the coarse water spray, which is injected radially into the venturi throat, into fine droplets. High energy and extreme turbulence promote collision between water droplets and dust particulates in the throat. The agglomeration process between particle and droplet continues in the diverging section of the venturi. The large agglomerates formed in the venturi are then removed by an inertial separator.

Venturi scrubbers achieve very high collection efficiencies for respirable dust. Since efficiency of a venturi scrubber depends on pressure drop, some manufacturers supply a variable-throat venturi to maintain pressure drop with varying gas flows.

Advantages and Disadvantages — Wet Scrubbers

Advantages	Disadvantages
• Have low capital costs and small space requirements	• Have high operating and maintenance costs
• Can treat high-temperature and high-humidity gas streams	• Require corrosion-resistant materials if used with acidic gases
• Are able to collect gases as well as particulates (especially "sticky" particulates)	• Require a precleaner for heavy dust loadings
• Have no secondary dust sources	• Cause water pollution; require further water treatment
	• Are susceptible to erosion at high velocities
	• Collect wet products
	• Require freeze protection

Common Operating Problems and Solutions — Wet Scrubbers

Problem	Solution
• Wet/dry buildup	• Keep all areas dry or all areas flooded. • Use inclined ducts to a liquid drain vessel. • Ensure that scrubber is installed vertically. • Maintain liquid seal.
• Dust buildup in fan	• Install clean water spray at fan inlet.
• Excessive fan vibration	• Clean fan housing and blades regularly.
• Liquid pump failure	• Divert some of the recycle slurry to a thickener, settling pond, or waste disposal area and supply clean water as makeup. • Increase the water bleed rate.
• Worn valves	• Use wear-resistant orifice plates to reduce erosion on valve components.
• Jammed valves	• Provide continuous purge between valves and operating manifold to prevent material buildup.
• Erosion of slurry piping	• Maintain pumping velocity of 4 to 6 ft/s to minimize abrasion and prevent sedimentation and settling.
• Plugged nozzles	• Replace nozzles or rebuild heads. • Change source of scrubbing liquid. • Supply filtered scrubbing liquid.
• Buildup on mist eliminators	• For vane-type demisters, spray the center and periphery intermittently to clean components. • For chevron-type demisters, spray the water from above to clean the buildup.

Startup/Shutdown Procedures — Wet Scrubbers

Prestart Checkout

1. Start fans and pumps to check their rotation.
2. Disconnect pump suction piping and flush it with water from an external source.
3. Install temporary strainers in pump suction line and begin liquid recycle.
4. With recycle flow on, set valves to determine operating conditions for desired flow rates. Record the valve positions as a future baseline.
5. Record all system pressure drops under clean conditions.
6. Perform all recommended lubrications.
7. Shut down fan, drain the system, and remove temporary strainers.

Startup

1. Allow vessels to fill with liquid through normal level controls. Fill large-volume basins from external sources.
2. Start liquid flow to all pump glands and fan sprays.
3. Start recycle pumps with liquid bleed closed.
4. Check insulation dampers and place scrubber in series with primary operation.
5. Start fan and fan inlet spray. Leave inlet control damper closed for 2 min to allow fan to reach speed.
6. Check gas saturation, liquid flows, liquid levels, fan pressure drop, duct pressure drops, and scrubber pressure drop.
7. Open bleed to pond, thickener, or other drain systems so slurry concentration can build slowly. Check final concentration as cross-check on bleed rate.

Shutdown

1. Shut down fan and fan spray. Insulate scrubber from operation.
2. Allow liquid system to operate as long as possible to cool and reduce liquid slurry concentrations.
3. Shut off makeup water and allow to bleed normally.
4. When pump cavitation noise is heard, turn off pump and pump gland water.
5. Open system manholes, bleeds, and other drains.

Preventive Maintenance Procedures — Wet Scrubbers

Frequency	Procedure
Daily	• Check recycle flow. • Check bleed flow. • Measure temperature rise across motor. • Check fan and pump bearings every 8 hours for oil level, oil color, oil temperature, and vibration. • Check scrubber pressure drop. • Check pump discharge pressure. • Check fan inlet and outlet pressure. • Check slurry bleed concentration. • Check vibration of fan for buildup or bleeds. • Record inlet and saturation temperature of gas stream. • Use motor current readings to detect flow decreases. Use fan current to indicate gas flow. • Check pressure drop across mesh and baffle mist eliminators. Clean by high-pressure spraying, if necessary.
Weekly	• Check wet/dry line areas for material buildup. Clean, if necessary. • Check liquid spray quantity and manifold pressure on mist eliminator automatic washdown. • Inspect fans on dirty applications for corrosion, abrasion, and particulate buildup. • Check bearings, drive mechanisms, temperature rise, sprocket alignment, sprocket wear, chain tension, oil level, and clarifier rakes. • Check ductwork for leakage and excessive flexing. Line or replace as necessary. • Clean and dry pneumatic lines associated with monitoring instrumentation.
Semiannually	• Verify accuracy of instruments and calibrate. • Inspect orifice plates. • Clean electrical equipment, including contacts, transformer insulation, and cooling fans. • Check and repair wear zones in scrubbers, valves, piping, and ductwork. • Lubricate damper drive mechanisms and bearings. Verify proper operation of dampers and inspect for leakage.

Electrostatic precipitators use electrostatic forces to separate dust particles from exhaust gases. A number of high-voltage, direct-current discharge electrodes are placed between grounded collecting electrodes. The contaminated gases flow through the passage formed by the discharge and collecting electrodes.

The airborne particles receive a negative charge as they pass through the ionized field between the electrodes. These charged particles are then attracted to a grounded or positively charged electrode and adhere to it.

The collected material on the electrodes is removed by rapping or vibrating the collecting electrodes either continuously or at a predetermined interval. Cleaning a precipitator can usually be done without interrupting the airflow.

The four main components of all electrostatic precipitators are—

● Power supply unit, to provide high-voltage, unidirectional current

● Ionizing section, to impart a charge to particulates in the gas stream

● A means of removing the collected particulates

● A housing to enclose the precipitator zone

The following factors affect the efficiency of electrostatic precipitators:

● Larger collection-surface areas and lower gas-flow rates increase efficiency because of the increased time available for electrical activity to treat the dust particles.

● An increase in the dust-particle migration velocity to the collecting electrodes increases efficiency. The migration velocity can be increased by—

-   Decreasing the gas viscosity
-   Increasing the gas temperature
-   Increasing the voltage field

Electrostatic Precipitators

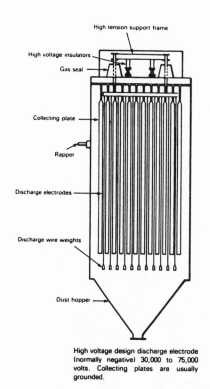

High voltage design discharge electrode (normally negative) 30,000 to 75,000 volts. Collecting plates are usually grounded.

Electrostatic Precipitator

Types of Precipitators

There are two main types of precipitators:

- **High-Voltage, Single-Stage** — Single-stage precipitators combine an ionization and a collection step. They are commonly referred to as Cottrell precipitators.

- **Low-Voltage, Two-Stage** — Two-stage precipitators use a similar principle; however, the ionizing section is followed by collection plates.

Described below is the high-voltage, single-stage precipitator, which is widely used in minerals processing operations. The low-voltage, two-stage precipitator is generally used for filtration in air-conditioning systems.

High-Voltage, Single-Stage Precipitators

The two major types of high-voltage precipitators currently used are—

- Plate
- Tubular

Plate Precipitators — The majority of electrostatic precipitators installed are the plate type. Particles are collected on flat, parallel surfaces that are 8 to 12 in. apart, with a series of discharge electrodes spaced along the centerline of two adjacent plates. The contaminated gases pass through the passage between the plates, and the particles become charged and adhere to the collection plates. Collected particles are usually removed by rapping the plates and deposited in bins or hoppers at the base of the precipitator.

Tubular Precipitators — Tubular precipitators consist of cylindrical collection electrodes with discharge electrodes located on the axis of the cylinder. The contaminated gases flow around the discharge electrode and up through the inside of the cylinders. The charged particles are collected on the grounded walls of the cylinder. The collected dust is removed from the bottom of the cylinder.

Tubular precipitators are often used for mist or fog collection or for adhesive, sticky, radioactive, or extremely toxic materials.

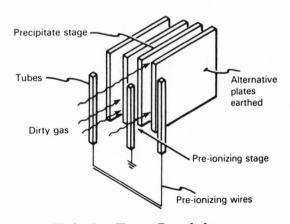

Plate-Type Precipitator

Tubular-Type Precipitator

Advantages and Disadvantages — Electrostatic Precipitators

Advantages	Disadvantages
• Have collection efficiencies in excess of 99% for all particulates, including sub-micron-sized particles	• Have high initial investment costs
• Usually collect dust by dry methods	• Do not respond well to process changes such as changes in gas temperature, gas pressure, gas flow rate, gaseous or chemical composition, dust loading, particulate size distribution, or electrical conductivity of the dust
• Have lower pressure drop and therefore lower operating costs	
• Can operate at high temperatures (up to 1200° F) and in colder climates	• Have a risk of explosion when gas stream contains combustibles
• Can remove acids and tars (sticky dust) as well as corrosive materials	• Produce ozone during gas ionization
• Allow increase in collection efficiency by increasing precipitator size	• Require large space for high efficiency, and even larger space for dust with low or high resistivity characteristics
• Require little power	• Require special precautions to protect personnel from exposure to high-voltage
• Can effectively handle relatively large gas flows (up to 2,000,000 ft^3/min)	• Require highly skilled maintenance personnel

Unit Collectors

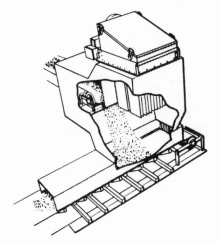

Unit Collector

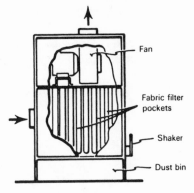

Fabric Collector

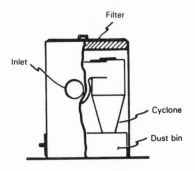

Cyclone Collector

Unlike central collectors, unit collectors control contamination at its source. They are small and self-contained, consisting of a fan and some form of dust collector. They are suitable for isolated, portable, or frequently moved dust-producing operations, such as bins and silos or remote belt-conveyor transfer points. Advantages of unit collectors include small space requirements, the return of collected dust to main material flow, and low initial cost. However, their dust-holding and storage capacities, servicing facilities, and maintenance periods have been sacrificed.

A number of designs are available, with capacities ranging from 200 to 2,000 ft^3/min. There are two main types of unit collectors:

- Fabric collectors, with manual shaking or pulse-jet cleaning — normally used for fine dust

- Cyclone collectors — normally used for coarse dust

Fabric collectors are frequently used in minerals processing operations because they provide high collection efficiency and uninterrupted exhaust airflow between cleaning cycles. Cyclone collectors are used when coarser dust is generated, as in woodworking, metal grinding, or machining.

The following points should be considered when selecting a unit collector:

- Cleaning efficiency must comply with all applicable regulations.

- The unit should maintain its rated capacity while accumulating large amounts of dust between cleanings.

- The cleaning operation should be simple and should not increase the surrounding dust concentration.

- The unit should be capable of operating unattended for extended periods of time (for example, 8 hours).

- The unit should have an automatic discharge or sufficient dust storage space to hold at least 1 week's accumulation.

- If renewable filters are used, they should not have to be replaced more than once a month.

- The unit should be durable.

- The unit should be quiet.

Use of unit collectors may not be appropriate if the dust-producing operations are located in an area where central exhaust systems would be practical. Dust-removal and servicing requirements are expensive for many unit collectors and are more likely to be neglected than those for a single, large collector.

Selecting a Dust Collector

Dust collectors vary widely in design, operation, effectiveness, space requirements, construction, and capital, operating, and maintenance costs. Each type has advantages and disadvantages. However, the selection of a dust collector should be based on the following general factors:

- **Dust Concentration and Particle Size** — For minerals processing operations, the dust concentration can range from 0.1 to 5.0 grains of dust per cubic feet of air, and the particle size can vary from 0.5 to 100 μm.

- **Degree of Dust Collection Required** — The degree of dust collection required depends on its potential as a health hazard or public nuisance, the plant location, the allowable emission rate, the nature of the dust, its salvage value, and so forth. The selection of a collector should be based on the efficiency required and should consider the need for high-efficiency, high-cost equipment, such as electrostatic precipitators; high-efficiency, moderate-cost equipment, such as baghouses or wet scrubbers; or lower cost, primary units, such as dry centrifugal collectors.

- **Characteristics of Airstream** — The characteristics of the airstream can have a significant impact on collector selection. For example, cotton fabric filters cannot be used where air temperatures exceed 180° F. Also, condensation of steam or water vapor can blind bags. Various chemicals can attack fabric or metal and cause corrosion in wet scrubbers.

● **Characteristics of Dust** — Moderate to heavy concentrations of many dusts (such as dust from silica sand or metal ores) can be abrasive to dry centrifugal collectors. Hygroscopic material can blind bag collectors. Sticky material can adhere to collector elements and plug passages. Some particle sizes and shapes may rule out certain types of fabric collectors. The combustible nature of many fine materials rules out the use of of electrostatic precipitators.

● **Methods of Disposal** — Methods of dust removal and disposal vary with the material, plant process, volume, and type of collector used. Collectors can unload continuously or in batches. Dry materials can create secondary dust problems during unloading and disposal that do not occur with wet collectors. Disposal of wet slurry or sludge can be an additional material-handling problem; sewer or water pollution problems can result if wastewater is not treated properly.

A comparison of various dust collector characteristics is included in the following table.

Comparison of Dust Collector Characteristics

Device	To Control Particulates Greater Than (μm)	Pressure Drop (in. wg)	Water Usage (gal/min per 1,000 ft³/min)	Humid Air Influence	Space Requirements	Maximum Temperature (1) (°F)	Costs (ft³/min) (2)
Cyclone	20-40	0.75-1.5	—	May cause condensation and plugging	Large	750	5¢-25¢
Multiclone	10-30	3-6	—	May cause condensation and plugging	Moderate	750	5¢-25¢
Shaker baghouse	0.25	3-6	—	May make bag cleaning difficult	Large	180 (3)	30¢-$2.50
Reverse-air baghouse	0.25	3-8	—		Moderate	550 (4)	
Reverse-jet baghouse	0.25	3-8	—	May cause bag to blind	Large	180 (3)	
Low-energy scrubber (e.g., spray tower)	25	0.5-2.5	5	None	Large	Unlimited	
Low- to medium-energy scrubber (e.g., centrifugal collector)	1-5	2.5-6	3-5	None	Moderate	Unlimited	25¢-75¢
Medium- to high-energy scrubber (e.g., packed bed)	1-5	6-15	5-10	None	Large	Unlimited	
High-energy scrubber (e.g., venturi)	0.5-2	15 and greater	5-15	None	Moderate	Unlimited	
Precipitator (single- or double-stage)	0.25	0.5	—	Improves efficiency	Large	500	50¢-$1.00

Notes: (1) Based on standard construction.
        (2) Cost based on collector section only.
             Does not include ducting, water, and power requirements.
             Cost figures should be used for comparison only. Actual costs may vary.
        (3) 180° F based on cotton fabric. Synthetic fabrics may be used to 275° F.
        (4) 550° F based on glass-fiber bags.

Fan and Motor

The fan and motor system supplies mechanical energy to move contaminated air from the dust-producing source to a dust collector.

Types of Fans

There are two main kinds of industrial fans:

- Centrifugal fans
- Axial-flow fans

Centrifugal Fans

Centrifugal fans consist of a wheel or a rotor mounted on a shaft that rotates in a scroll-shaped housing. Air enters at the eye of the rotor, makes a right-angle turn, and is forced through the blades of the rotor by centrifugal force into the scroll-shaped housing. The centrifugal force imparts static pressure to the air. The diverging shape of the scroll also converts a portion of the velocity pressure into static pressure.

There are three main types of centrifugal fans:

- **Radial-Blade Fans** — Radial-blade fans are used for heavy dust loads. Their straight, radial blades do not get clogged with material, and they withstand considerable abrasion. These fans have medium tip speeds and medium noise factors.

- **Backward-Blade Fans** — Backward-blade fans operate at higher tip speeds and thus are more efficient. Since material may build up on the blades, these fans should be used after a dust collector. Although they are noisier than radial-blade fans, backward-blade fans are commonly used for large-volume dust collection systems because of their higher efficiency.

- **Forward-Curved-Blade Fans** — These fans have curved blades that are tipped in the direction of rotation. They have low space requirements, low tip speeds, and a low noise factor. They are usually used against low to moderate static pressures.

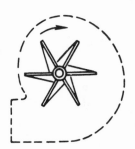

Radial Blades

Backward Blades

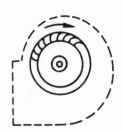

Forward-Curved Blades

Axial-Flow Fans

Axial-flow fans are used in systems that have low resistance levels. These fans move the air parallel to the fan's axis of rotation. The screw-like action of the propellers moves the air in a straight-through parallel path, causing a helical flow pattern.

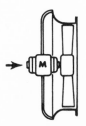

Propeller Fan

The three main kinds of axial fans are—

- **Propeller Fans** — These fans are used to move large quantities of air against very low static pressures. They are usually used for general ventilation or dilution ventilation and are good in developing up to 0.5 in. wg.

- **Tube-Axial Fans** — Tube-axial fans are similar to propeller fans except they are mounted in a tube or cylinder. Therefore, they are more efficient than propeller fans and can develop up to 3 to 4 in. wg. They are best suited for moving air containing substances such as condensable fumes or pigments.

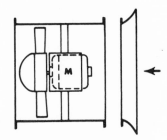

Tube-Axial Fan

- **Vane-Axial Fans** — Vane-axial fans are similar to tube-axial fans except air-straightening vanes are installed on the suction or discharge side of the rotor. They are easily adapted to multistaging and can develop static pressures as high as 14 to 16 in. wg. They are normally used for clean air only.

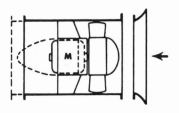

Vane-Axial Fan

When selecting a fan, the following points should be considered:

- Volume required

- Fan static pressure

- Type of material to be handled through the fan (For example, a radial-blade fan should be used with fibrous material or heavy dust loads, and nonsparking construction must be used with explosive or inflammable materials.)

- Type of drive arrangement, such as direct drive or belt drive

- Space requirements

Fan Selection

Direct Driven	Belt Driven
• Less space requirements	• Greater space requirements
• Assure constant fan speed	• Fan speeds easily changed (a vital factor in many applications)
• Fan speeds limited to available motor speeds	

- Noise levels

- Operating temperature (For example, sleeve bearings are suitable to 250° F; ball bearings to 550° F.)

- Sufficient size to handle the required volume and pressure with minimum horsepower

- Need for special coatings or construction when operating in corrosive atmospheres

- Ability of fan to accommodate small changes in total pressure while maintaining the necessary air volume

- Need for an outlet damper to control airflow during cold starts (If necessary, the damper may be interlocked with the fan for a gradual start until steady-state conditions are reached.)

Fan Rating Tables

After the above information is collected, the actual selection of fan size and speed is usually made from a rating table published by the fan manufacturer. This table is known as a multirating table, and it shows the complete range of capacities for a particular size of fan.

Points to Note:

- The multirating table shows the range of pressures and speeds possible within the limits of the fan's construction.

- A particular fan may be available in different construction classes (identified as class I through IV) relating to its capabilities and limits.

- For a given pressure, the highest mechanical efficiency is usually found in the middle third of the volume column.

- A fan operating at a given speed can have an infinite number of ratings (pressure and volume) along the length of its characteristic curve. However, when the fan is installed in a dust collection system, the point of rating can only be at the point at which the system resistance curve intersects the fan characteristic curve.

- In a given system, a fan at a fixed speed or at a fixed blade setting can have a single rating only. This rating can be changed only by changing the fan speed, blade setting, or the system resistance.

- For a given system, an increase in exhaust volume will result in increases in static and total pressures. For example, for a 20% increase in exhaust volume in a system with 5 in. pressure loss, the new pressure loss will be $5 \times (1.20)^2 = 7.2$ in.

- For rapid estimates of probable exhaust volumes available for a given motor size, the equation for brake horsepower, as illustrated, can be useful.

Fan ratings for volume and static pressure, as described in the multirating tables, are based on the tests conducted under ideal conditions. Often, field installation creates airflow problems that reduce the fan's air delivery. The following points should be considered when installing the fan:

- Avoid installation of elbows or bends at the fan discharge, which will lower fan performance by increasing the system's resistance.

- Avoid installing fittings that may cause nonuniform flow, such as an elbow, mitred elbow, or square duct.

- Check that the fan impeller is rotating in the proper direction—clockwise or counterclockwise.

- For belt-driven fans—
  - Check that the motor sheave and fan sheave are aligned properly.
  - Check for proper belt tension.

- Check the passages between inlets, impeller blades, and inside of housing for buildup of dirt, obstructions, or trapped foreign matter.

Brake Horsepower Equation

$$bhp = \frac{cfm \times TP}{6356 \times ME \text{ of fan}}$$

where:

cfm = air volume, ft^3/min

TP = total pressure, inches of water

ME = mechanical efficiency of fan (Operating points will be 0.50 to 0.65 for most centrifugal fans.)

Fan Installation

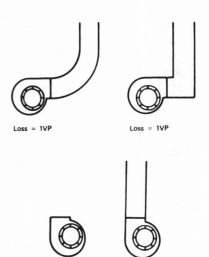

Loss = 1VP   Loss = 1VP

Loss = 0.5VP   No loss

**Typical
Fan Discharge Conditions**

Electric Motors

Electric motors are used to supply the necessary energy to drive the fan. They are normally classified in two groups:

- Integrals
- Fractionals

Integral-horsepower electric motors are normally three-phase, alternating-current motors. Fractional-horsepower electric motors are normally single-phase, alternating-current motors and are used when less than 1 hp is required. Since most dust collection systems require motors with more than 1 hp, only integral-horsepower motors are discussed here.

Types of Motors

The two most common types of integral-horsepower motors used in dust collection systems are—

- **Squirrel-Cage Motors** — These motors have a constant speed and are of a nonsynchronous, induction type.

- **Wound-Rotor Motors** — These motors are also known as slip-ring motors. They are general-purpose or continuous-rated motors and are chiefly used when an adjustable-speed motor is desired.

Squirrel-cage and wound-rotor motors are further classified according to the type of enclosure they use to protect their interior windings. These enclosures fall into two broad categories:

- Open
- Totally enclosed

Drip-proof and splash-proof motors are open motors. They provide varying degrees of protection; however, they should not be used where the air contains substances that might be harmful to the interior of the motor.

Totally enclosed motors are weather-protected with the windings enclosed. These enclosures prevent free exchange of air between the inside and the outside, but they are not airtight.

Totally enclosed, fan-cooled (TEFC) motors are another kind of totally enclosed motor. These motors are the most commonly used motors in dust collection systems. They have an integral-cooling fan outside the enclosure, but within the protective shield, that directs air over the enclosure.

Both open and totally-enclosed motors are available in explosion-proof and dust-ignition-proof models to protect against explosion and fire in hazardous environments.

Selection of Motor

When selecting a motor, the following points should be considered:

- Required brake horsepower and revolutions per minute

- Characteristics of power supply, such as line voltage (110, 220, 440 volts), single-phase or three-phase alternating current, and frequency

- Environmental conditions under which motor would be operated (for example, temperature, humidity, corrosive atmospheres, or open flames)

- Characteristics of the load (i.e., the fan and drive elements) and power company restrictions on starting current

- Necessary overload protection for the motor

- Ability to supply adequate power under "cold" starts

Fan Troubleshooting Chart

Symptom	Probable Cause	Solution
Insufficient airflow, low ft³/min	**Fan**	
	• Forward curved impeller installed backwards	• Reinstall impeller
	• Fan running backwards	• Change fan rotation by reversing two of the three leads on the motor
	• Impeller not centered with inlet collar(s)	• Make impeller and inlet collar(s) concentric
	• Fan speed too low	• Increase fan speed by installing smaller diameter pulley
	• Elbows or other obstructions restricting airflow	• Redesign ductwork • Install turning vanes in elbow • Remove obstructions in ductwork
	• No straight duct at fan inlet	• Install straight length of ductwork, at least 4 to 6 duct diameters long, where possible • Increase fan speed to overcome this pressure loss
	• Obstruction near fan outlet	• Remove obstruction or redesign ductwork near fan outlet
	• Sharp elbows near fan outlet	• Install a long radius elbow, if possible • Install turning vanes in elbow
	• Improperly designed turning vanes	• Redesign turning vanes
	• Projections, dampers, or other obstructions near fan outlet	• Remove all obstructions

Symptom	Probable Cause	Solution
	Duct System	
	● Actual system more restrictive (more resistant to flow) than expected	● Decrease system's resistance by redesigning ductwork
	● Dampers closed	● Open or adjust all dampers according to the design
	● Leaks in supply ducts	● Repair all leaks in supply duct
Too much airflow, high ft³/min	**Fan**	
	● Backward inclined impeller installed backwards (high horsepower)	● Install impeller as recommended by manufacturer
	● Fan speed too fast	● Reduce fan speed
		● Install larger diameter pulley on fan
	Duct System	
	● Oversized ductwork; less resistance	● Redesign ductwork or add restrictions to increase resistance
	● Access door open	● Close all access and inspection doors
Low static pressure, high ft³/min	**Fan**	
	● Backward inclined impeller installed backwards (high horsepower)	● Install impeller as recommended by manufacturer
	● Fan speed too high	● Reduce fan speed
		● Install larger diameter pulley on fan
	Duct System	
	● System has less resistance to flow than expected	● Reduce fan speed to obtain desired flow rate

Symptom	Probable Cause	Solution
Low static pressure, low ft³/min	**Gas Density** • Gas density lower than anticipated (due to high temperature gases or high altitudes)	• Calculate gas flow rate at desired operating conditions by applying appropriate correction factors for high temperature or altitude conditions
	Duct System • Fan inlet and/or outlet conditions not same as tested	• Increase fan speed • Install smaller diameter pulley on fan • Redesign ductwork
High static pressure, low ft³/min	**Duct System** • Obstructions in system	• Remove obstructions
	• Duct system too restricted	• Redesign ductwork • Install larger diameter ducts
High horsepower	**Fan** • Backward inclined impeller installed backwards	• Install impeller as recommended by manufacturer
	• Fan speed too high	• Reduce fan speed • Install larger diameter pulley on fan
	Duct System • Oversized ductwork	• Redesign ductwork
	• Access door open	• Close all access/inspection doors

Symptom	Probable Cause	Solution
	Gas Density	
	• Calculated horsepower requirements based on light gas (e.g., high temperature or high altitude) but actual gas is heavy (e.g., cold startup)	• Replace motor • Install outlet damper, which will open gradually until fan comes to its operating speed
	Fan Selection	
	• Fan not operating at efficient point of rating	• Redesign system • Change fan • Change motor • Consult fan manufacturer
Fan does not operate	**Electrical**	
	• Blown fuses	• Replace blown fuses
	• Electricity turned off	• Turn on electricity
	• Wrong voltage	• Check for proper voltage on fan
	• Motor too small and overload protector has broken circuit	• Change motor to a larger size
	Mechanical	
	• Broken belts	• Replace belts
	• Loose pulleys	• Tighten or reinstall pulleys
	• Impeller touching scroll	• Reinstall impeller properly

Disposal of Collected Dust

After dust-laden exhaust gases are cleaned, the collected dust must be disposed of properly. Ideally, dust can be returned to the product stream and sold; if this is not possible, disposal of dust may become a problem. For example, when dry dust collectors are used, secondary dust problems may arise during unloading and disposal of collected dust; for wet dust collectors, the disposal of wet slurry or sludge can be a problem.

Proper disposal of collected dust can be accomplished in four steps:

1. Removing dust from the hopper of the dust collector

2. Conveying the dust

3. Storing the dust

4. Treating the dust for final disposal

Removing Dust from Hopper

Collected dust must be removed from the hopper to prevent recirculation within the dust collector. Since the material must be removed continuously (while the dust collector is operating), rotary air locks or tipping valves should be used to maintain a positive air seal. If the material in the hopper has a bridging tendency, equipment such as bin vibrators, rappers, or air jets should be used.

Conveying the Dust

After the dust has been removed from the collector, it must be transported to a central point for accumulation and ultimate disposal. Conveying of dust can be accomplished by the following methods:

- Use of screw conveyors

- Use of air conveyors (pneumatic conveying)

- Use of air slides (low-pressure pneumatic conveying)

- Use of pumps and piping systems to convey slurry

Screw conveyors have been used with a great deal of success. However, trouble areas to be considered are maintenance access, worn-out bearings and casings due to abrasive materials, and air leaks. For wet dust collectors, inclined conveyors can be used to convey the slurry to a settling pond.

Pneumatic conveyors are often selected to convey dry dust because they have few moving parts and can convey either horizontally or vertically. They operate on a high-velocity, low-air-volume principle. Trouble areas include excessive wear and abrasion in the piping and high capital and operating costs.

Air slides are commonly used for nonabrasive, light dust. They work on the principle of air fluidization of dust particles and are useful for heavy horizontal conveying. Trouble areas include ability to maintain a certain downward slope and greater maintenance requirements.

Pumps and piping systems are used to convey the slurry to a settling pond. However, care must be taken in this method to prevent water-pollution.

Storing the Dust

After the material has been removed and transported from the dust collector, a storage facility must be used to permit disposal in efficient quantities. Elevated storage tanks or silos are normally used to permit loading of dry dust into enclosed trucks underneath.

For wet dust collectors, the accumulation area is a settling pond. A settling pond may require considerable space. Since the storage area can only be decanted and dried out during the dry season, two settling ponds are usually needed. Also, most collected materials have very fine components that may seal the pond and prevent percolation.

Treating the Dust for Final Disposal

In most cases, the disposal of fine dust requires great care to prevent recirculation by the wind. Several final dust disposal methods commonly used are—

- Landfilling
- Recycling
- Pelletizing
- Byproduct utilization
- Backfilling mines and quarries

Chapter 5
Specific Illustrations

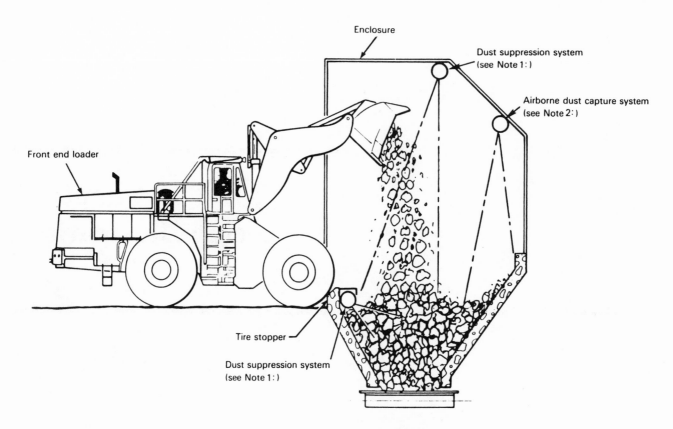

Truck or Loader to Crusher

Note 1 : Dust suppression system should be activated only during dumping cycle.

Note 2 : Airborne dust capture system should also be activated during dumping cycle and may be kept on beyond the dumping cycle until adequate control of dust is achieved.

Truck or Loader to Crusher

$$Q_E = 33.3 \times \left(\frac{600\,T}{G}\right)$$

Q_E = Exhaust air volume, CFM
T = Weight of material dumped, TPM, tons per minute
G = Bulk density of material, pounds/Ft^3

Note: The open area at the face of the enclosure should be kept to a minimum.

Dry

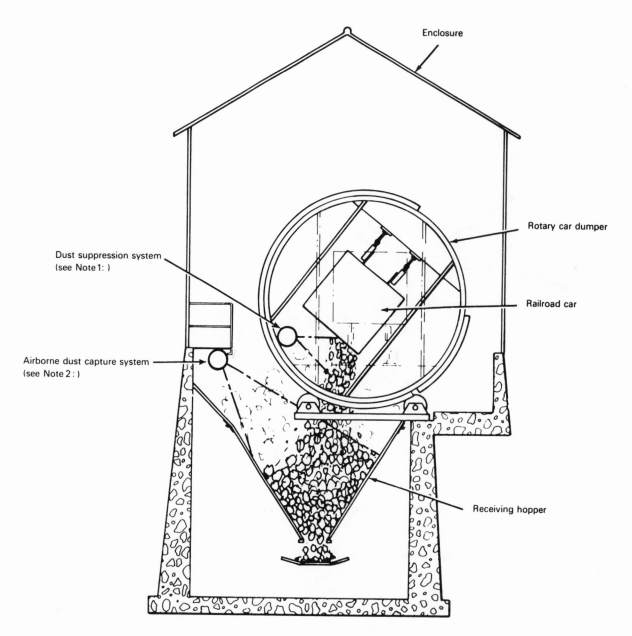

Railroad Rotary Dump to Receiving Hopper

Note 1 : Dust suppression system should be activated only during dumping cycle.

Note 2 : Airborne dust capture system should be activated during dumping cycle and may be kept on beyond the dumping cycle until adequate control of dust is achieved.

Wet

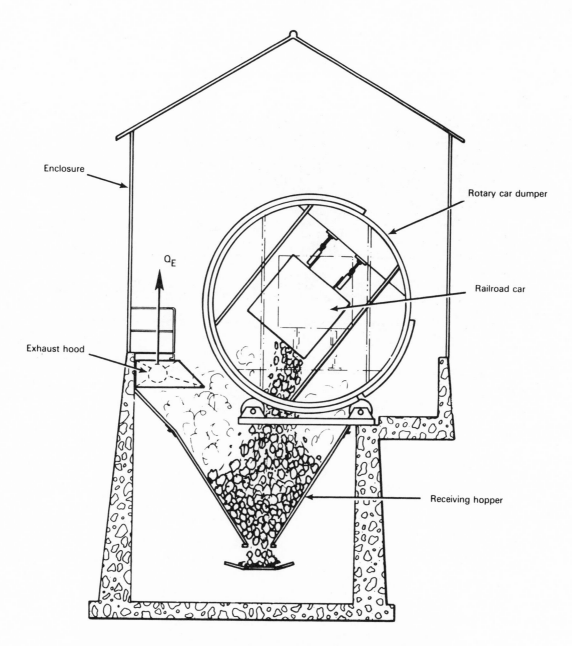

Enclosure

Rotary car dumper

Q_E

Railroad car

Exhaust hood

Receiving hopper

Railroad Rotary Dump to Receiving Hopper

$$Q_E = 33.3 \times \left(\frac{600\ T}{G}\right)$$

Q_E = Exhaust air volume, CFM
T = Weight of material dumped, TPM, tons per minute
G = Bulk density of material, pounds/Ft^3

Dry

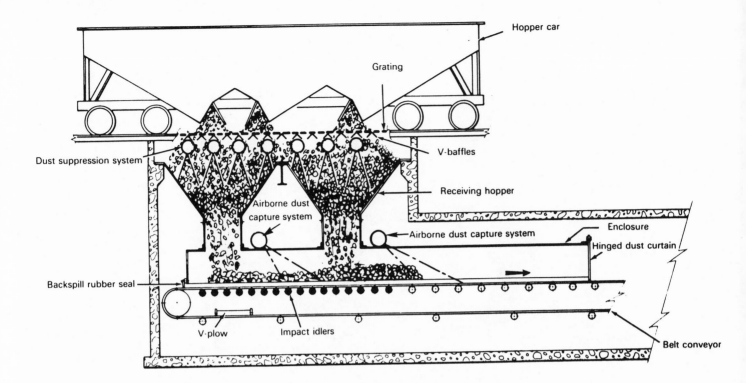

Railroad Dump to Conveyor

Wet

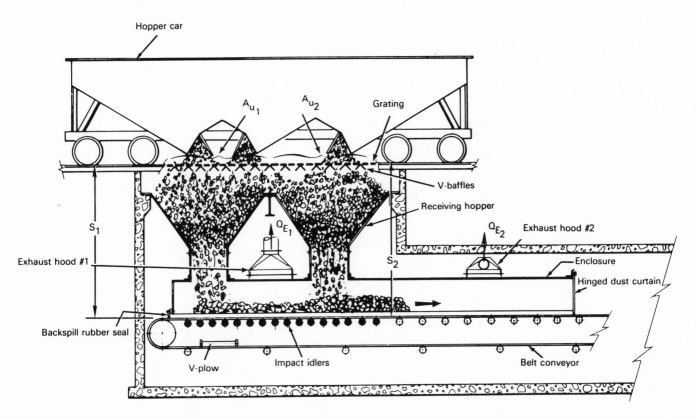

Railroad Dump to Conveyor

$$Q_{E_1} = 10\, A_{u_1} \sqrt{\frac{R_1\, S_1^{\,2}}{D}}$$

$$Q_{E_2} = 10\, A_{u_2} \sqrt{\frac{R_2\, S_2^{\,2}}{D}}$$

Q_{E_1}, Q_{E_2} = Exhaust air volume, CFM
A_{u_1}, A_{u_2} = Enclosure open area at upstream, FT^2
R_1, R_2 = Rate of material flow, TPM
S_1, S_2 = Height of fall, FT
D = Average material size, FT

Note 1 : Q_{E_1} and Q_{E_2} should be calculated by using appropriate values of A_u , R and S .

Note 2 : This equation should be applied where D is greater than 1/8 in.

Dry

Feeder to:

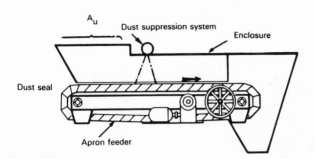

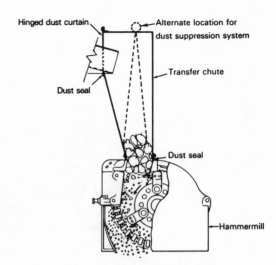

Hammermill Crusher

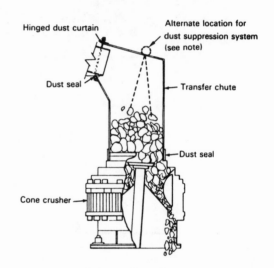

Cone Crusher

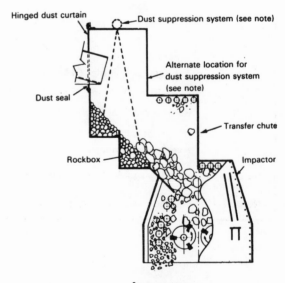

Impactor

Wet

Feeder to:

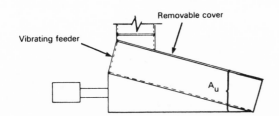

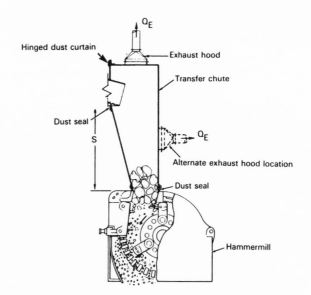

Hammermill Crusher

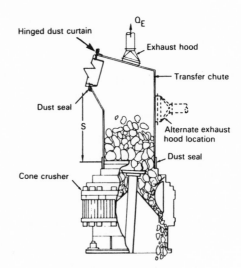

Cone Crusher

$$Q_E = 10 \times A_u \times \sqrt{\frac{RS^2}{D}}$$

Q_E = Exhaust air volume, CFM
A_u = Enclosure open area at upstream, FT^2
R = Rate of material flow, TPH
S = Height of fall, FT
D = Average material size, FT

Note: This equation should be applied where D is greater than 1/8 in.

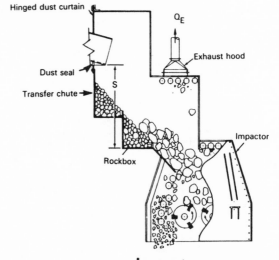

Impactor

Dry

Feeder to:

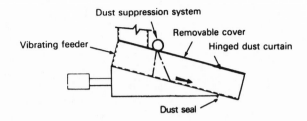

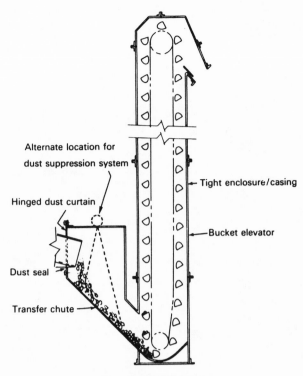

Bucket Elevator

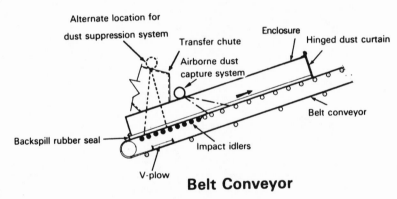

Belt Conveyor

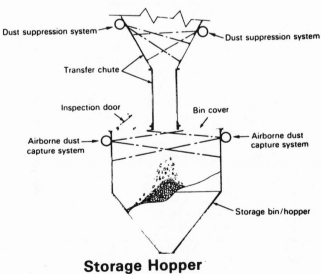

Storage Hopper

Wet

Feeder to:

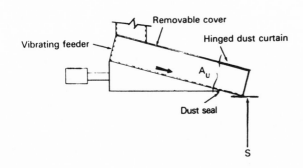

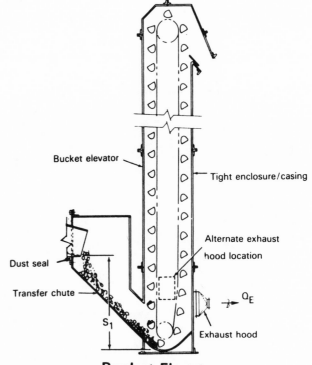

Bucket Elevator

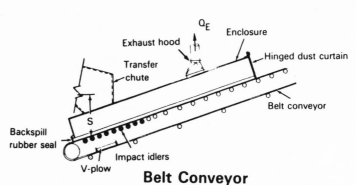

Belt Conveyor

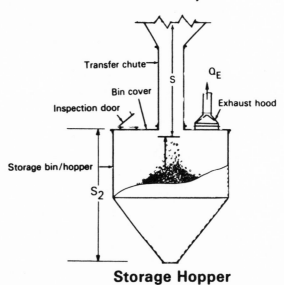

Storage Hopper

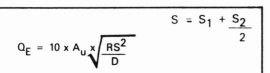

$$S = S_1 + \frac{S_2}{2}$$

$$Q_E = 10 \times A_u \times \sqrt{\frac{RS^2}{D}}$$

Q_E = Exhaust air volume, CFM
A_u = Enclosure open area at upstream, FT^2
R = Rate of material flow, TPH
S = Height of fall, FT
D = Average material size, FT

Note: This equation should be applied where
D is greater than 1/8 in.

Dry

Feeder to: Screen or Mill

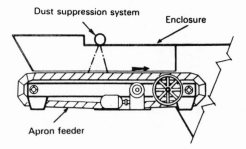

Not Recommended

Wet

Feeder to:

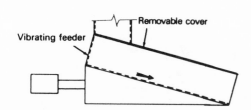

Removable cover

Vibrating feeder

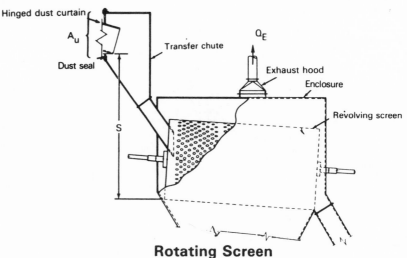

Hinged dust curtain

A_u

Dust seal

Transfer chute

Q_E

Exhaust hood

Enclosure

Revolving screen

S

Rotating Screen

Hinged dust curtain

A_u

Dust seal

Transfer chute

S

Q_E

Exhaust hood

Screen covers

Dust seal

Dust seals

Vibrating screen

Vibrating Screen

$$Q_E = 10 \times A_u \times \sqrt{\frac{RS^2}{D}}$$

Q_E = Exhaust air volume, CFM
A_u = Enclosure open area at upstream, FT^2
R = Rate of material flow, TPH
S = Height of fall, FT
D = Average material size, FT

Note: This equation should be applied where
D is greater than 1/8 in.

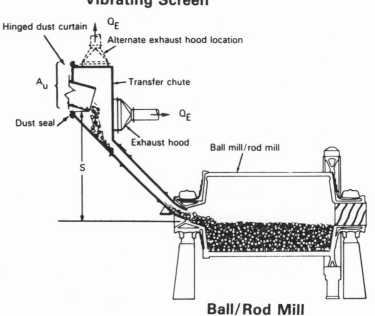

Hinged dust curtain

Q_E

Alternate exhaust hood location

A_u

Transfer chute

Q_E

Dust seal

Exhaust hood

Ball mill/rod mill

S

Ball/Rod Mill

Dry

Conveyor to:

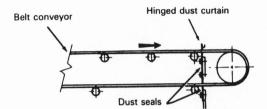

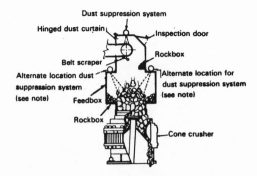

Cone Crusher

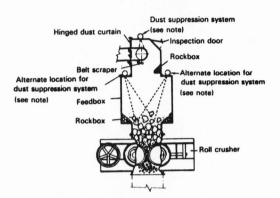

Roll Crusher

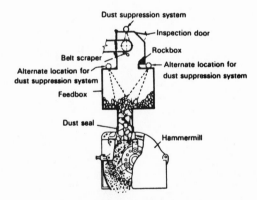

Hammermill Crusher

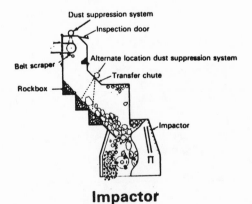

Impactor

Note: Excessive moisture may cause crusher jamming

Wet

Conveyor to:

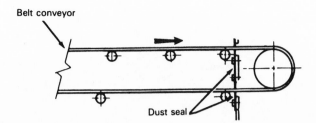

Belt conveyor

Dust seal

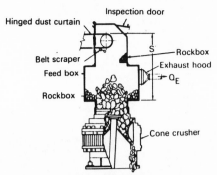

Cone Crusher

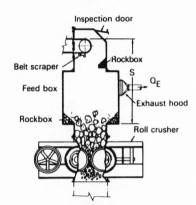

Roll Crusher

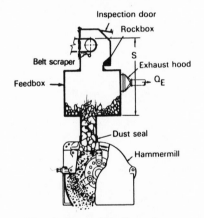

Hammermill Crusher

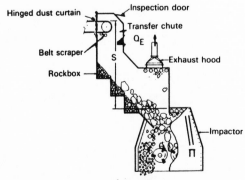

Impactor

$$Q_E = 10 \times A_u \times \sqrt{\frac{RS^2}{D}}$$

Q_E = Exhaust air volume, CFM
A_u = Enclosure open area at upstream, FT^2
R = Rate of material flow, TPH
S = Height of fall, FT
D = Average material size, FT

Note: This equation should be applied where
D is greater than 1/8 in.

Dry

Conveyor to:

Belt conveyor

Dust seal

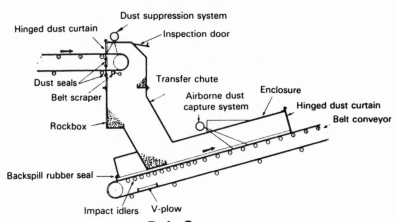

Dust suppression system

Inspection door

Hinged dust curtain

Bucket elevator

Tight enclosure/casing

Dust seal

Belt scraper

Rockbox

Mini rockbox

Transfer chute

Bucket Elevator

Dust suppression system

Hinged dust curtain

Inspection door

Transfer chute

Airborne dust capture system

Enclosure

Hinged dust curtain

Belt conveyor

Dust seals

Belt scraper

Rockbox

Backspill rubber seal

Impact idlers   V-plow

Belt Conveyor

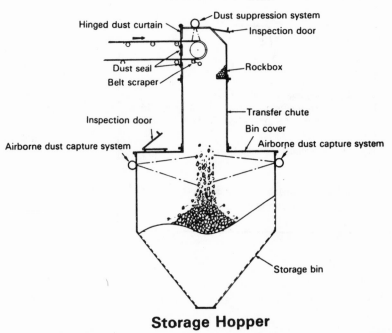

Dust suppression system

Hinged dust curtain

Inspection door

Dust seal

Rockbox

Belt scraper

Transfer chute

Inspection door

Bin cover

Airborne dust capture system

Airborne dust capture system

Storage bin

Storage Hopper

Wet

Conveyor to:

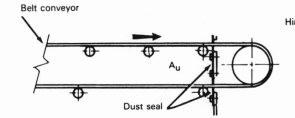

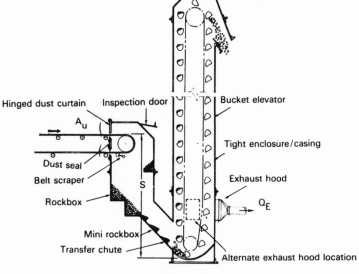

Bucket Elevator

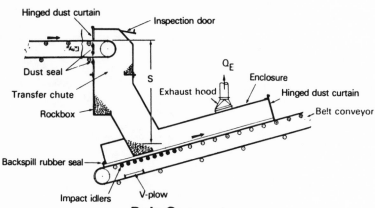

Belt Conveyor

$$S = S_1 + \frac{S_2}{2}$$

$$Q_E = 10 \times A_u \times \sqrt{\frac{RS^2}{D}}$$

Q_E = Exhaust air volume, CFM
A_u = Enclosure open area at upstream, FT^2
R = Rate of material flow, TPH
S = Height of fall, FT
D = Average material size, FT

Note: This equation should be applied where
   D is greater than 1/8 in.

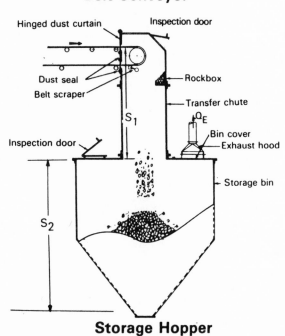

Storage Hopper

Dry

Conveyor to: Screen or Mill

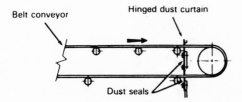

Not Recommended

Wet

Conveyor to:

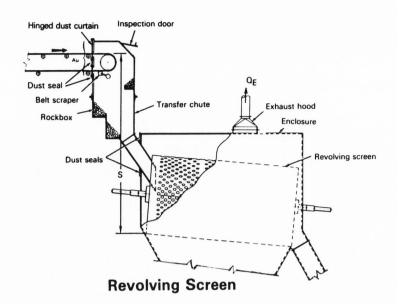

Revolving Screen

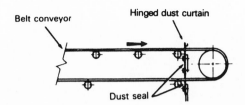

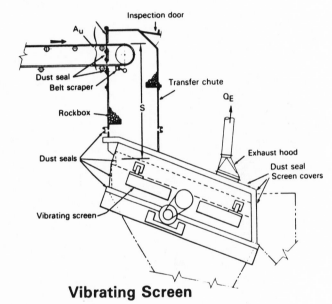

Vibrating Screen

$$Q_E = 10 \times A_u \times \sqrt{\frac{RS^2}{D}}$$

Q_E = Exhaust air volume, CFM
A_u = Enclosure open area at upstream, FT^2
R = Rate of material flow, TPH
S = Height of fall, FT
D = Average material size, FT

Note: This equation should be applied where
      D is greater than 1/8 in.

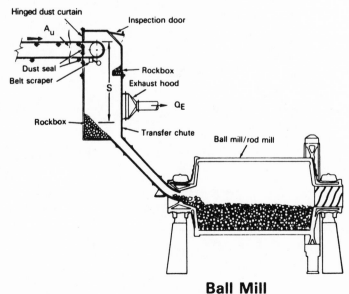

Ball Mill

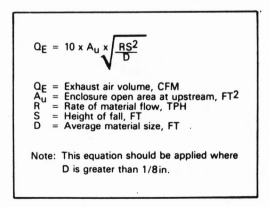

Dry

From: Jaw Crusher

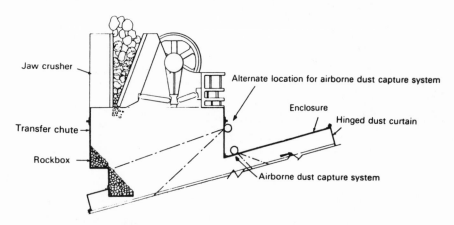

From: Roll Crusher

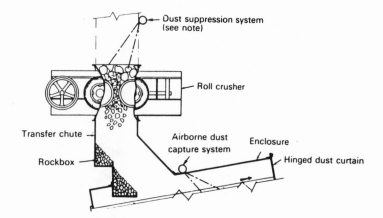

From: Impactor

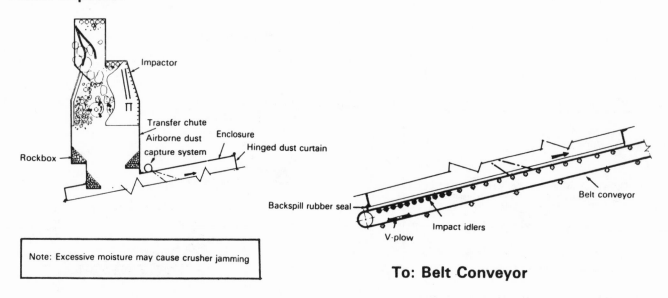

Note: Excessive moisture may cause crusher jamming

To: Belt Conveyor

Wet

From: Jaw Crusher

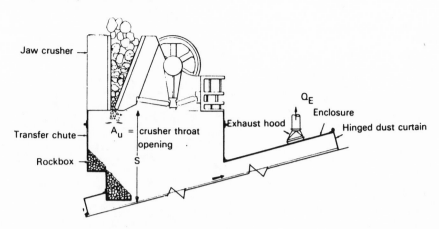

From: Roll Crusher

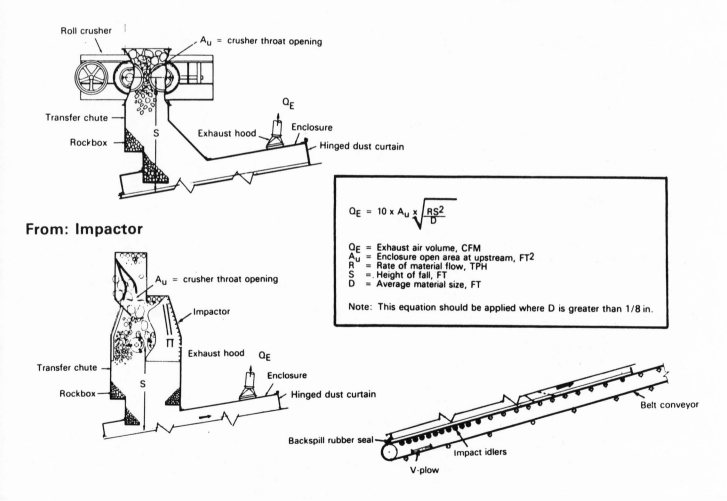

From: Impactor

$$Q_E = 10 \times A_u \times \sqrt{\frac{RS^2}{D}}$$

Q_E = Exhaust air volume, CFM
A_u = Enclosure open area at upstream, FT^2
R = Rate of material flow, TPH
S = Height of fall, FT
D = Average material size, FT

Note: This equation should be applied where D is greater than 1/8 in.

To: Belt Conveyor

Dry

From: Cone Crusher

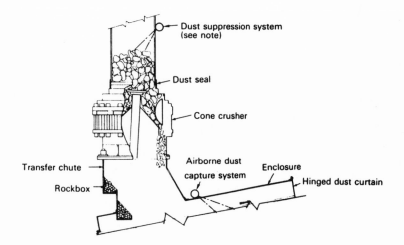

From: Hammermill Crusher

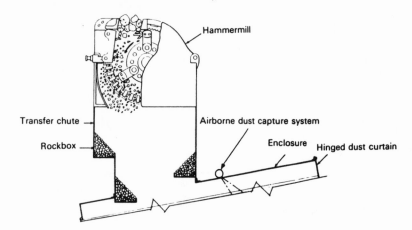

From: Gyratory Crusher

Note: Excessive moisture may cause crusher jamming

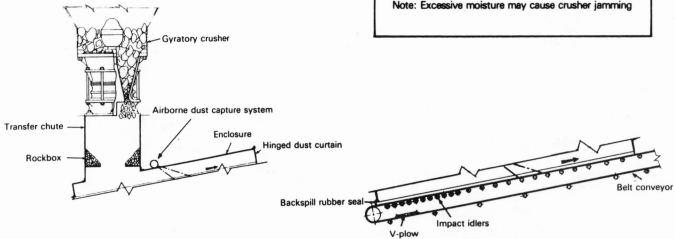

To: Belt Conveyor

Wet

From: Cone Crusher

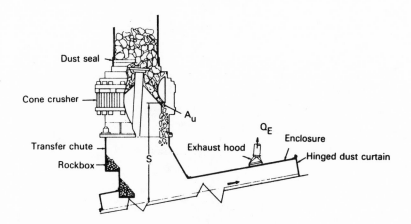

From: Hammermill Crusher

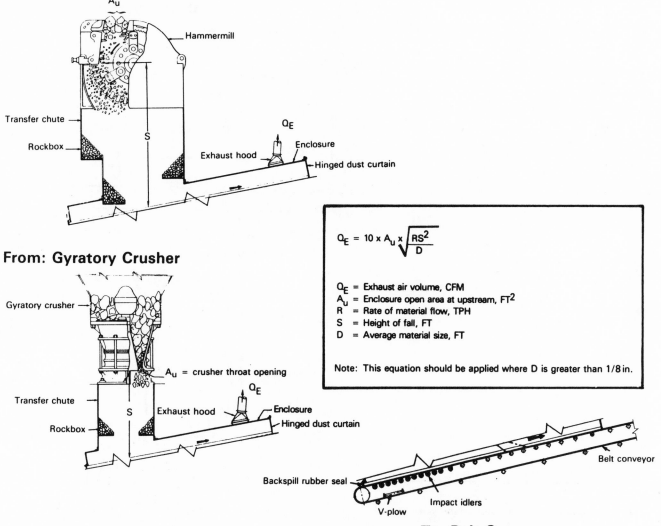

$$Q_E = 10 \times A_u \times \sqrt{\frac{RS^2}{D}}$$

Q_E = Exhaust air volume, CFM
A_u = Enclosure open area at upstream, FT^2
R = Rate of material flow, TPH
S = Height of fall, FT
D = Average material size, FT

Note: This equation should be applied where D is greater than 1/8 in.

From: Gyratory Crusher

A_u = crusher throat opening

To: Belt Conveyor

Dry

Screen to:

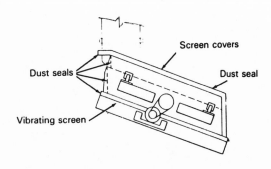

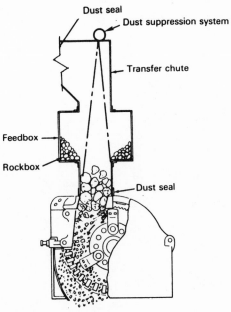

Crushers

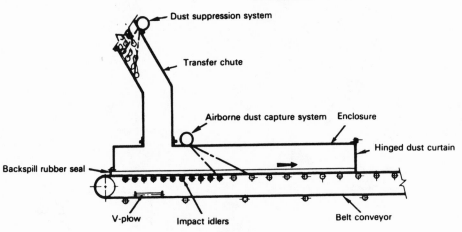

Conveyors

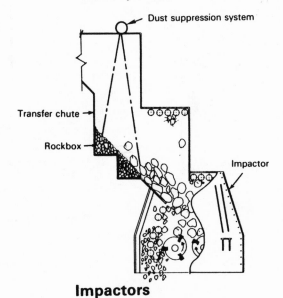

Impactors

Note: Excessive moisture may cause crusher jamming

Wet

Screen to:

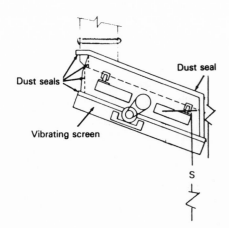

Vibrating screen · Dust seals · Dust seal · S

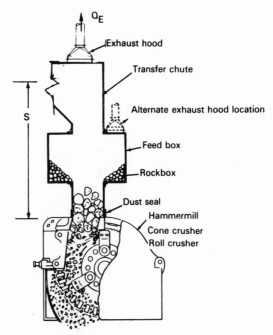

Crushers

Exhaust hood · Transfer chute · Alternate exhaust hood location · Feed box · Rockbox · Dust seal · Hammermill · Cone crusher · Roll crusher · Q_E · S

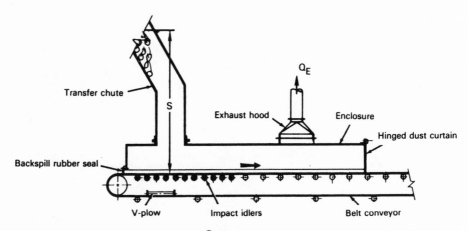

Conveyors

Transfer chute · Exhaust hood · Enclosure · Hinged dust curtain · Backspill rubber seal · V-plow · Impact idlers · Belt conveyor · Q_E · S

$$Q_E = 10 \times A_u \times \sqrt{\frac{RS^2}{D}}$$

Q_E = Exhaust air volume, CFM
A_u = Enclosure open area at upstream, FT^2
R = Rate of material flow, TPH
S = Height of fall, FT
D = Average material size, FT

Note: This equation should be applied where D is greater than 1/8 in.

Note: A_u depends upon tightness of enclosure at the screen

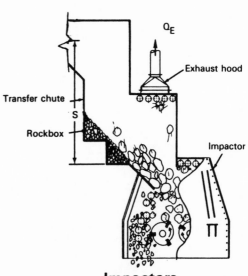

Impactors

Transfer chute · Rockbox · Exhaust hood · Impactor · Q_E · S

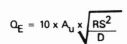

Dry

Screen to:

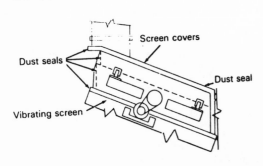

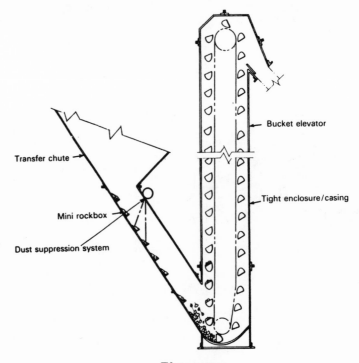

Elevators

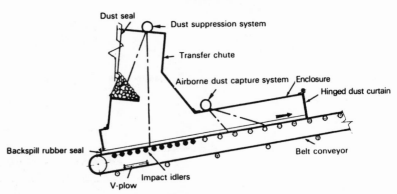

Conveyors

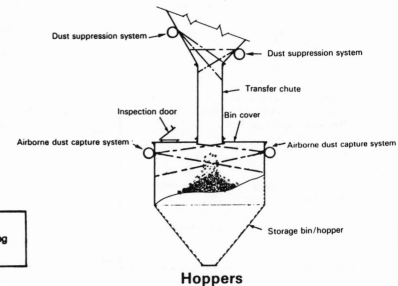

Hoppers

Note: Excessive moisture may cause crusher jamming

Wet

Screen to:

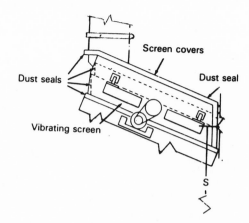

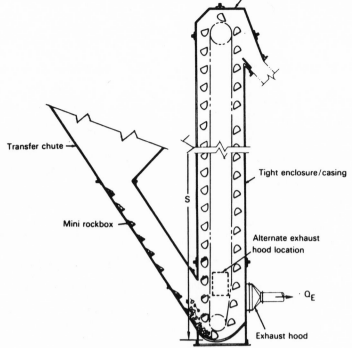

Elevators

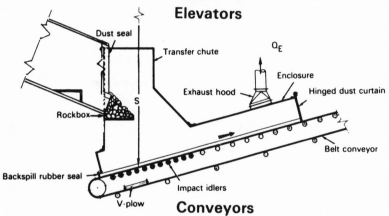

Conveyors

$$Q_E = 10 \times A_u \times \sqrt{\frac{RS^2}{D}}$$

$$S = S_1 + \frac{S_2}{2}$$

Q_E = Exhaust air volume, CFM
A_u = Enclosure open area at upstream, FT^2
R = Rate of material flow, TPH
S = Height of fall, FT
D = Average material size, FT

Note: This equation should be applied where
       D is greater than 1/8 in.

Note: "A_u" depends upon tightness of enclosure
       at the screen.

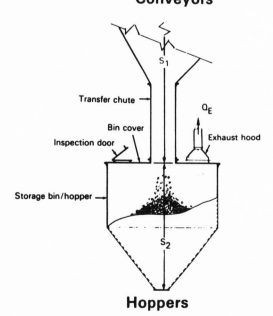

Hoppers

Dry

Screen to:

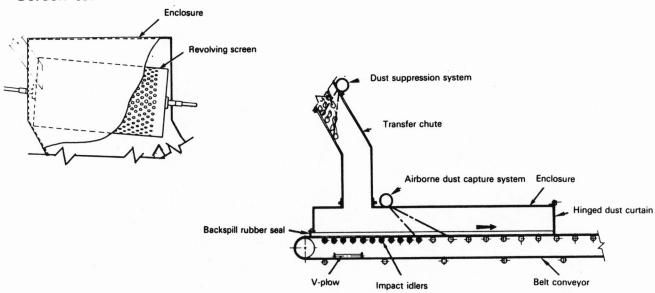

Conveyors

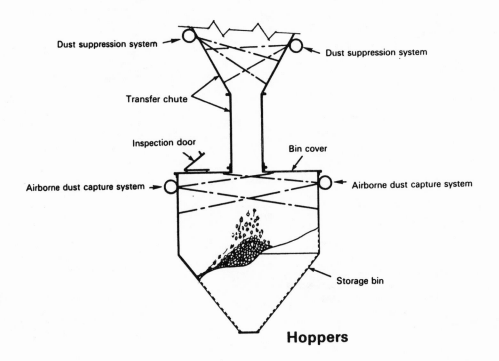

Hoppers

Wet

Screen to:

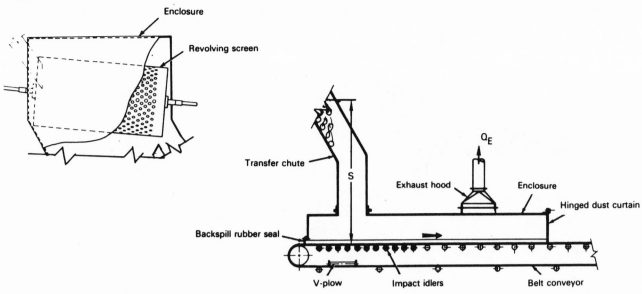

Conveyors

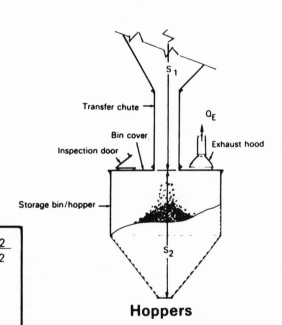

Hoppers

$$S = S_1 + \frac{S_2}{2}$$

$$Q_E = 10 \times A_u \times \sqrt{\frac{RS^2}{D}}$$

Q_E = Exhaust air volume, CFM
A_u = Enclosure open area at upstream, FT^2
R = Rate of material flow, TPH
S = Height of fall, FT
D = Average material size, FT

Note: This equation should be applied where
      D is greater than 1/8 in.

Note: A_u depends upon tightness of enclosure
      at the screen

Dry

Ball/Rod Mill to: Feeder or Screen

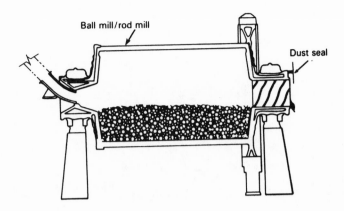

Not Recommended

Wet

Ball/Rod Mill to:

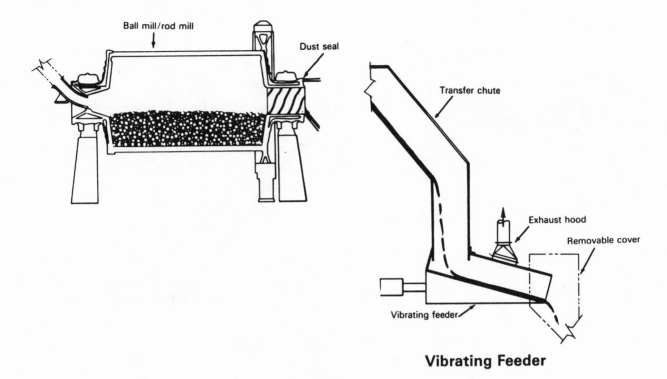

Vibrating Feeder

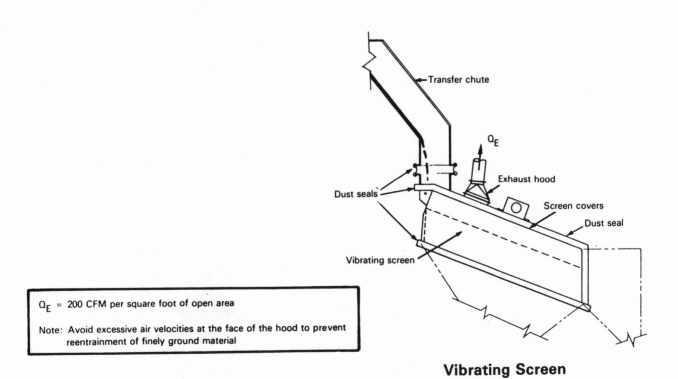

Q_E = 200 CFM per square foot of open area

Note: Avoid excessive air velocities at the face of the hood to prevent reentrainment of finely ground material

Vibrating Screen

Dry

Ball/Rod Mill to:

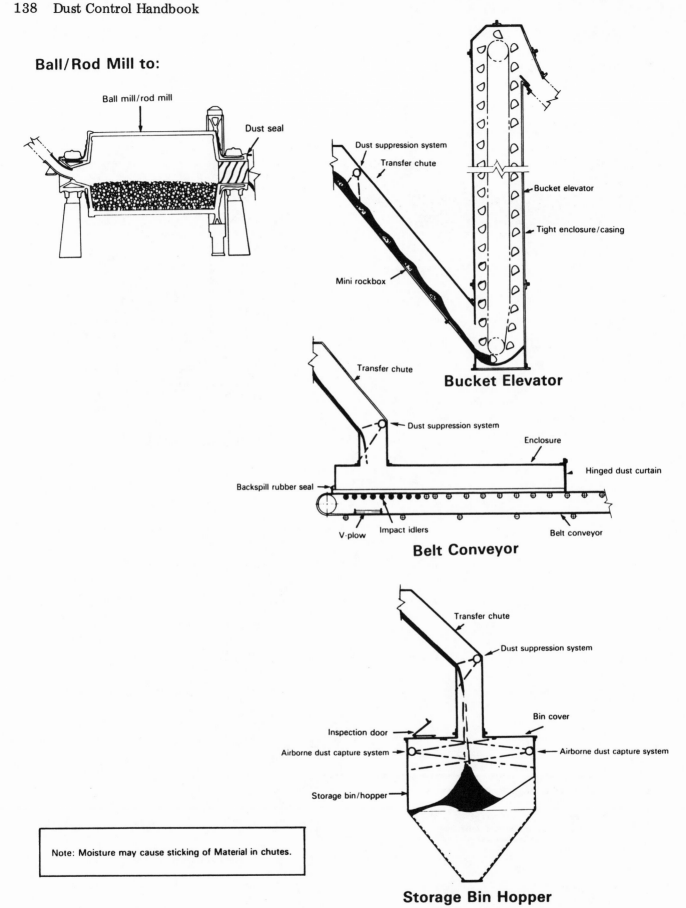

Ball mill/rod mill

Dust seal

Dust suppression system

Transfer chute

Bucket elevator

Tight enclosure/casing

Mini rockbox

Bucket Elevator

Transfer chute

Dust suppression system

Enclosure

Hinged dust curtain

Backspill rubber seal

V-plow

Impact idlers

Belt conveyor

Belt Conveyor

Transfer chute

Dust suppression system

Bin cover

Inspection door

Airborne dust capture system

Airborne dust capture system

Storage bin/hopper

Note: Moisture may cause sticking of Material in chutes.

Storage Bin Hopper

Wet

Ball/Rod Mill to:

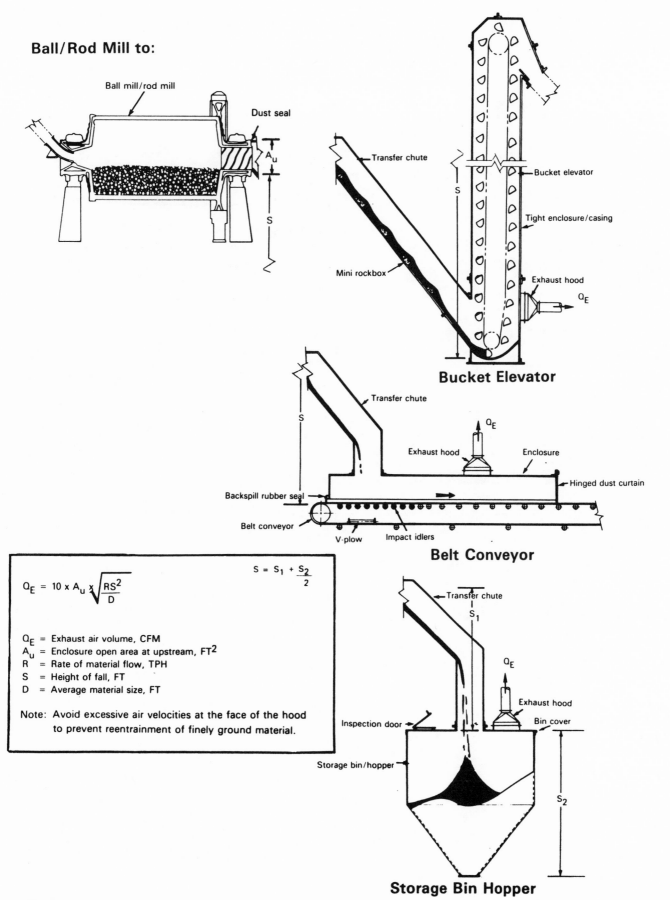

Bucket Elevator

Belt Conveyor

Storage Bin Hopper

$$Q_E = 10 \times A_u \sqrt{\frac{RS^2}{D}}$$

$$S = S_1 + \frac{S_2}{2}$$

Q_E = Exhaust air volume, CFM
A_u = Enclosure open area at upstream, FT^2
R = Rate of material flow, TPH
S = Height of fall, FT
D = Average material size, FT

Note: Avoid excessive air velocities at the face of the hood
       to prevent reentrainment of finely ground material.

Dry

Elevator to:

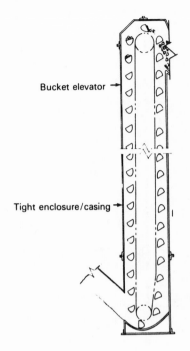

Bucket elevator

Tight enclosure/casing

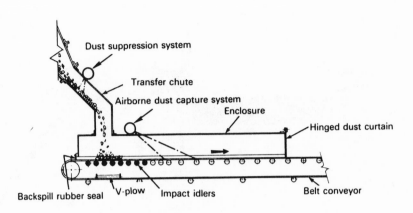

Belt Conveyor

Note: Wet systems are not recomended for fine powdery material

Elevator to: Screens

Not Recommended

Wet

Elevator to:

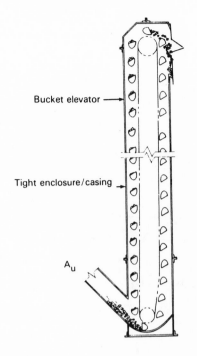

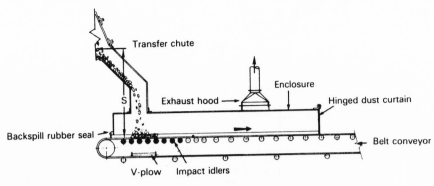

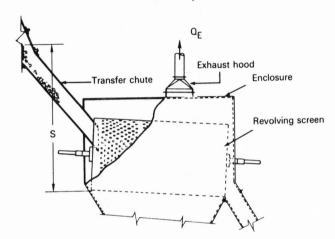

Belt Conveyor

Revolving Screen

$$Q_E = 10 \times A_u \times \sqrt{\frac{RS^2}{D}}$$

Q_E = Exhaust air volume, CFM
A_u = Enclosure open area at upstream, FT^2
R = Rate of material flow, TPH
S = Height of fall, FT
D = Average material size, FT

Note: This equation should be applied where
D is greater than 1/8 in.

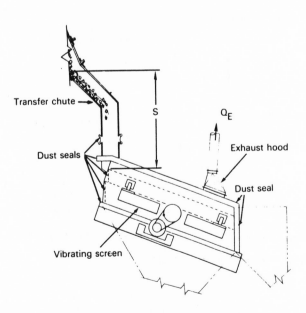

Vibrating Screen

Dry

Elevator to:

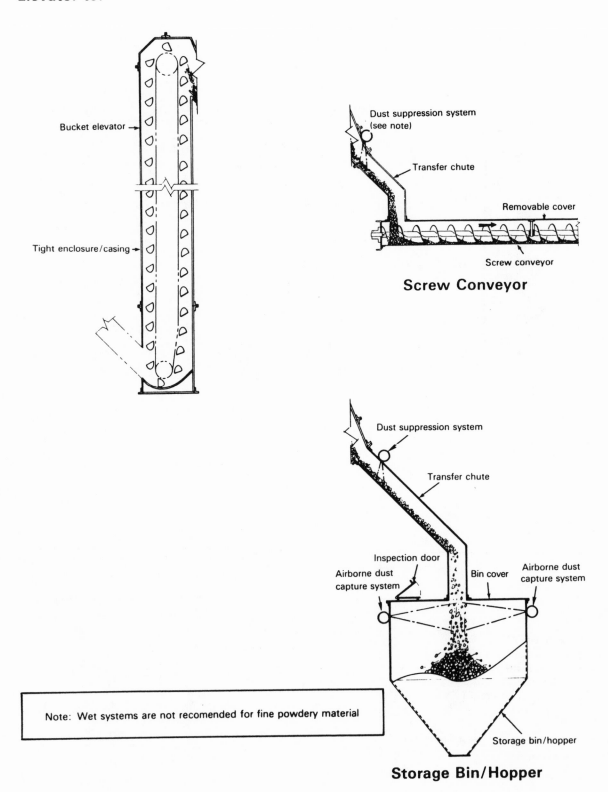

Bucket elevator →

Tight enclosure/casing →

Dust suppression system
(see note)

Transfer chute

Removable cover

Screw conveyor

Screw Conveyor

Dust suppression system

Transfer chute

Inspection door

Airborne dust
capture system

Bin cover

Airborne dust
capture system

Storage bin/hopper

Storage Bin/Hopper

Note: Wet systems are not recomended for fine powdery material

Wet

Elevator to:

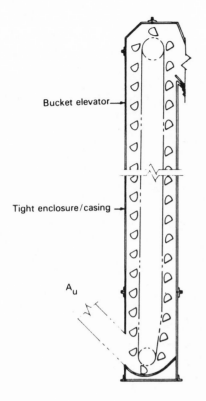

Bucket elevator

Tight enclosure/casing

A_u

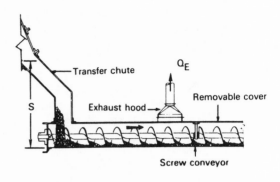

Transfer chute

Q_E

Exhaust hood

Removable cover

Screw conveyor

Screw Conveyor

$$Q_E = 10 \times A_u \times \sqrt{\frac{RS^2}{D}}$$

$$S = S_1 + \frac{S_2}{2}$$

Q_E = Exhaust air volume, CFM
A_u = Enclosure open area at upstream, FT^2
R = Rate of material flow, TPH
S = Height of fall, FT
D = Average material size, FT

Note: This equation should be applied where
D is greater than 1/8 in.

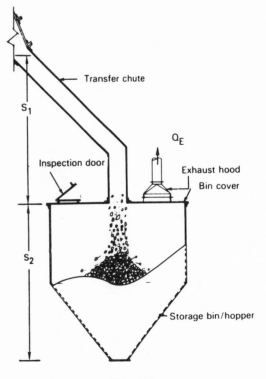

Transfer chute

S_1

Inspection door

Q_E

Exhaust hood

Bin cover

S_2

Storage bin/hopper

Storage Bin/Hopper

Dry

Hopper to:

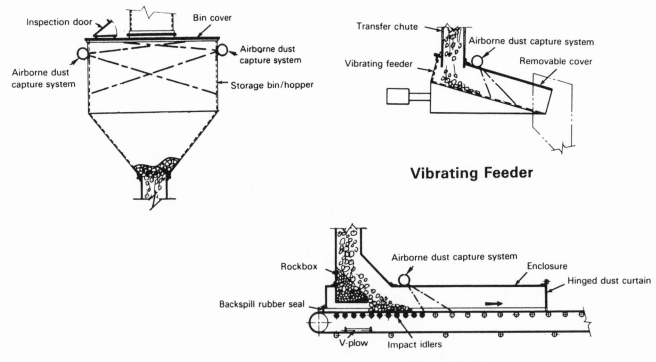

Vibrating Feeder

Belt Conveyor

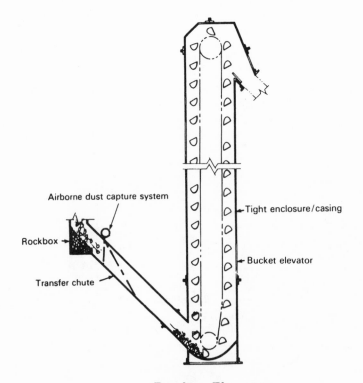

Bucket Elevator

Wet

Hopper to:

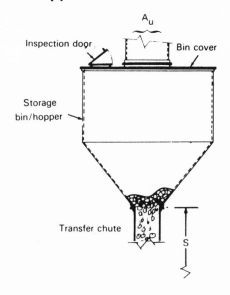

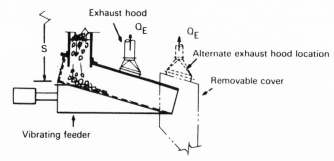

Vibrating Feeder

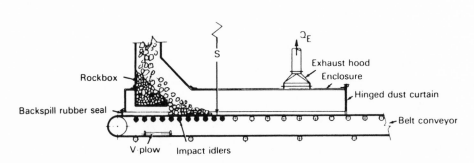

Belt Conveyor

$$Q_E = 10 \times A_u \times \sqrt{\frac{RS^2}{D}}$$

Q_E = Exhaust air volume, CFM
A_u = Enclosure open area at upstream, FT^2
R = Rate of material flow, TPH
S = Height of fall, FT
D = Average material size, FT

Note: This equation should be applied where
D is greater than 1/8 in.

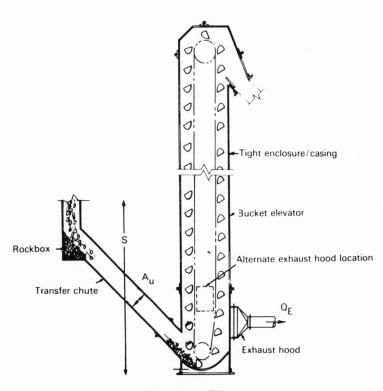

Bucket Elevator

Dry

Chapter 6

Estimating Costs of Dust Control Systems

Since adequate control of dust emissions can usually be achieved by more than one dust control method, a considerable economic burden may result if the appropriate method is not selected. This burden can be higher capital costs, higher operating and maintenance costs, or both. Preliminary cost estimates can prevent this unnecessary economic burden by—

- Characterizing the costs to be incurred and determining whether they are feasible

- Comparing expected costs of alternative dust control techniques to help identify the optimum control technique

Necessity for Cost Estimates

Several methods, with varying degrees of accuracy, are available for estimating costs. They range from presenting an overall installed cost on a per-unit basis to presenting detailed cost estimates based on preliminary designs, schematics, and/or vendors' cost estimates. The least accurate method is to equate overall costs to a basic operating parameter such as tons per hour or cubic feet per minute. This approach is not recommended. Where possible, detailed cost estimates should be arrived at by preparing preliminary designs and schematics. However, if time does not permit this approach, equipment vendors may be contacted. Based on their current knowledge of the technology and experience in the industry, they can provide reasonably accurate cost estimates.

Cost Estimating Procedures

Cost Components

To prepare and analyze cost estimates, a basic knowledge of the cost components and their relationship to the total cost of the system is essential.

Total costs for any system can be divided into—

- Capital costs
- Operating and maintenance costs

Capital Costs

Capital costs consist of the delivered costs for major control equipment, auxiliary equipment and accessories, and field installation. Capital costs can be grouped as follows:

Summary of Capital Costs

- Major control equipment (35%)*
- Auxiliary or accessory equipment (15%)
- Field installation (20%)
- Project management and engineering (13%)
- Freight, taxes, subcontractor, and so forth (17%) and
- Startup cost, working capital, and other capitalized costs (15-20%)**

---

* Average percent of the capital investment

** Additional costs, expressed as a percentage of total capital costs

- **Major Control Equipment**
  - Baghouses
  - Electrostatic precipitators
  - Scrubbers
  - Cyclones
  - Water-spray bars
  - Nozzles

- **Auxiliary or Accessory Equipment**
  - Air-moving equipment
    - Fans and blowers
    - Electrical motors, starters, wire, conduit, switches, etc.
    - Hoods, ductwork, gaskets, dampers, etc.

  - Liquid-moving equipment
    - Pumps
    - Compressors
    - Electrical motors, starters, wire, conduit, switches, etc.
    - Piping and valves
    - Settling tanks (for wet scrubbers)

  - Storage and disposal equipment
    - Dust storage bins and hoppers
    - Sludge pits
    - Drag lines, trackway, etc.

  - Supporting construction
    - Structural steelwork
    - Cement foundation, piers, etc.
    - Thermal insulation
    - Vibration and antiwear materials
    - Protective covers

- Instruments to measure or control the following:
  - Air and liquid flow
  - Temperature and pressure
  - Operation and capacity
  - Power
  - Opacity of flue gas
  - Dust concentration

● **Field Installation Costs**

- Labor required for delivery, assembly, removal or relocation of equipment
- Freight, taxes, and subcontractors' fees
- Engineering supervision
- Startup and performance testing
- Extending or increasing utilities

Baghouses

The capital costs of a baghouse depend on the following:

● Type of filtering media used (cotton, dacron, glass, Teflon, etc.)

● Type of fabric used (felted vs. woven)

● Adopted air-to-cloth ratios

● Type of cleaning mechanism (mechanical shaker, pulse jet, reverse air)

● Type of baghouse (suction type vs. pressure type; continuous duty vs. intermittent)

● Design and construction (standard design vs. custom design; carbon steel vs. stainless steel)

● Temperature of exhaust gases (high temperature vs. low ambient temperature)

Electrostatic Precipitators

The capital costs of an electrostatic precipitator depend on the net plate area (NPA). The NPA, in turn, depends on the efficiency required of the precipitator.

Following are the factors that affect the cost of an electrostatic precipitator:

- The electrical characteristics of the dust may have a significant effect on the collection efficiency and the plate area.

- The resistivity of the dust varies with the temperature and moisture content of the bag. Therefore, in some cases, auxiliary equipment may be required to precondition the gas stream before it enters the precipitator.

- The addition of moisture to the gas stream, in combination with a low operating temperature, may require insulating the precipitator to prevent condensation and corrosion.

Scrubbers

The capital costs of a scrubber depend, generally, on the following three factors:

- Volumetric airflow rate
- Operating pressure
- Construction

The volumetric airflow rate is the most important factor because the size of the scrubber and its cost are determined by the actual gas volume at the scrubber's inlet.

Operating pressure also affects scrubber efficiency and price. The higher the air volume and/or operating pressure, the greater the plate thickness of the shell must be to prevent buckling of the shell.

The cost of a scrubber can also increase if it is constructed of special materials, such as stainless steel or fiber-reinforced plastics to protect against corrosion or lining the scrubber shell with PVC, rubber, or refractories to protect against erosion.

Cyclones/Multiclones

The capital costs of a cyclone or a multiclone are a function of the particulate-removal efficiency, which, in turn, depends on the inlet gas velocity and inlet diameter. Theoretically, the higher the velocity or the smaller the inlet diameter, the greater the efficiency and pressure drop.

The material of construction also affects the cost. For handling highly abrasive dust, the cyclone/multiclone may have to be constructed of abrasion-resistant material or lined with ceramic material. For a highly corrosive gas stream, stainless steel or fiber-reinforced plastic may be necessary.

Fans and Motors

The capital costs of a fan are based on—

- Construction
- Class
- Volume handled
- Pressure developed

The capital costs of a fan motor are related to—

- Fan speed
- Total system pressure
- Gas volume flow rate
- Selected motor housing

Pumps

Although pump prices vary with the design of the pump, they are generally a function of—

- Pump head developed, ft
- Pump capacity, gal/min
- Pump speed, r/min

The selection of revolutions per minute for these pumps should be based on the design flow rate:

Flow Rate (gal/min)	Pump (r/min)
0 - 1,000	3,550
500 - 5,000	1,750
2,000 - 10,000	1,170

Note: Generally, the capital cost of the pump and motor combination varies inversely with the revolutions per minute; however, maintenance costs may be higher as the revolutions per minute increase.

Operating and Maintenance Costs

Operating and maintenance costs consist of direct expenses of labor and materials for operating and maintenance, the cost of replacement parts, utility costs, and waste disposal costs. They may also include indirect costs of overhead, taxes, insurance, general administration, and capital recovery charges. However, only direct costs are discussed here.

Operating Costs

Operating costs include—

● Direct labor and materials

● Utilities

Direct Labor and Materials Costs — Labor and materials costs for operation and maintenance of dust control systems vary substantially among plants due to factors such as the degree of automation, equipment age, and operating periods. Generally speaking, labor costs can be reduced by increased system automation.

For small- to medium-size systems with an installed cost of approximately $100,000 or less, the total cost of maintenance is approximately 5% of the installed cost.

Estimated Labor Hours per Shift

Control Device	Operating Labor (man-hours/shift)	Maintenance Labor (man-hours/shift)
Cyclone	0.5-2	1-2
Fabric filters/baghouses	2-4	1-2
Electrostatic precipitators	0.5-2	0.5-1
Scrubbers	2-8	1-2
Water spray system/ wet dust suppression system	2-4	1-2

Notes:

● Estimates are based on large plants operating three shifts per day for 365 days. For smaller plants expected to operate one shift per day, 5 days per week, 50 weeks per year, the labor hours/shift will be higher.

● Estimates are only for performing preventive maintenance.

● Where periodic replacement of major parts are required, such as replacement of filter bags in a baghouse or replacement of spray nozzles in a wet dust suppression system, the labor cost of replacement will be at least equal to the material cost of replacement parts.

Guidelines for Parts and Equipment Life

	Low (years)	Average (years)	High (years)
Materials and Parts Life			
Filter bags	0.3	1.5	5
Spray nozzles	0.01	0.5-1.5	2-3
Equipment Life			
Cyclone	5	20	40
Fabric filters	5	20	40
Electrostatic precipitators	5	20	40
Venturi scrubbers	5	10	20

Notes:

● The guidelines for average life represent a process operating continuously with three shifts per day, 5-7 days per week, 52 weeks per year.

● The guidelines for low life are based on a continuous process, handling moderate- to high-temperature gas streams, with a high concentration of corrosive or abrasive dusts.

● Applications having high life expectations for parts and equipment are assumed to be operating intermittently or approximately one shift per day with gas streams having ambient temperature and low dust concentrations.

Utility Costs — The utility costs for equipment such as pumps and electrical motors are a function of power/energy requirements, which can be calculated as follows:

Fan Power

$$kW \cdot h = 0.746 \, (hp) \, (H)$$

$$= \frac{0.746 \, (ft^3/min)(\Delta P)(SG)(H)}{6356 \, \eta}$$

where:

kW·h	= kilowatt hour
hp	= horsepower
ft^3/min	= actual volumetric airflow rate
P	= pressure loss, in. wg
	= mechanical efficiency, usually 60–70%
H	= hours of operation
SG	= specific gravity as compared to air at 70° F, 29.92 inches of mercury

Pump Power

$$kW \cdot h = 0.746 \, (hp) \, (H)$$

$$= \frac{0.746 \, (gal/min)(h_d)(SG)(H)}{3960}$$

where:

gal/min	= flow rate
hp	= horsepower
H	= hours of operation
h_d	= head of fluid (ft.)
SG	= specific gravity relative to water

Baghouse Power (auxiliaries, motor, etc.)

Horsepower requirements for baghouse shaker motors, reverse-air fan motors, etc. can be estimated at approximately 0.5 hp per 1,000 ft^2 of cloth area. Power usage depends on dust loading and cleaning frequency. Assuming a 50%

usage factor, power requirements are approximately 0.2 kW·h for 1,000 ft^2 of cloth area.

Electrostatic Precipitator Power

The power requirements for an electrostatic precipitator are approximately 1.5 W/ft^2 of collection plate area. The range varies from 0.3 to 3 W/ft^2.

Once the power requirement is known, the annual power costs can be calculated using the following equation:

Annual power cost ($) =

$$\begin{array}{ccccc} \text{Power usage} & & \text{Cost of power} & & \text{Total annual} \\ \text{(kW·h)} & \times & \text{($/kW·h)} & \times & \text{operating hrs.} \end{array}$$

Waste Disposal Costs

The cost of waste disposal includes the removal and hauling of dry contaminants to a nearby site. This cost varies with the particular plant and available landfill site.

Water Costs

Water costs vary in different areas.

Maintenance Costs

Maintenance costs include—

● Labor and materials for preventive and routine maintenance, such as lubrication, surface protection, cleaning, and painting

● Replacement of worn-out equipment, parts, or structures due to wear, abrasion, or corrosion

The annual cost of replacement parts represents the cost of the parts or components divided by their expected life. Replacement parts are components such as filter bags and spray nozzles, which have a limited life and must be replaced periodically.

Cost Justification

A 10-15% return on investment is necessary to justify any capital investment. However, when dust controls are considered, such a return is not always practical. The following are some tangible benefits that may assist in economic justification of a dust control system:

- Industrial taxes amount to about 50% of net income, which means business investments result in a tax savings equivalent to approximately half the expenditure.

- Return on investment before taxes is approximately twice the value of the return calculated after taxes.

Various federal and state governments provide the following tax relief benefits for industries that install dust control or pollution control systems:

- Section 169 of the Internal Revenue Tax Reform Act of 1969 permits a faster tax write-off of pollution control facilities. If the facility is certified by both the Environmental Protection Agency and the appropriate state agencies, facilities installed after 1968 can be amortized over a 60-month period. An exception to this is any control system that recovers the costs by generating a profit in some manner.

- Thirty-seven states provide tax incentives for pollution control facilities, such as—

  - Property tax exemption
  - Sales and use tax exemption
  - Income tax credit
  - Accelerated amortization

- All but four states (California, Idaho, New Jersey, and Texas) authorize the use of industrial revenue bonds to finance pollution control equipment. These are normally 15-year bonds that provide tax-free interest to the holders. Although the interest rates offered by these bonds are about 2% lower than most other bonds, they can attract investors because of their tax-free status.

Other intangible benefits may further assist in justifying a dust control system:

- Reducing health hazard possibilities

- Reducing risk of dust explosion and fire

- Reducing equipment wear and damage

- Increasing visibility

- Reducing or eliminating unpleasant odors

- Improving relations with neighbors

- Creating safer and more pleasant working conditions, thus improving employee morale and productivity

- Possible product or byproduct savings

Chapter 7
Controlling Surrounding Dust Sources

Introduction

Although a dust control system is the best way to reduce dust levels in a workplace, it cannot control dust from secondary sources such as material spillage or leakage from a process or piece of equipment, or dust brought in through the ventilation system or through doors and windows. These secondary sources can contribute significantly to dust levels and can render the existing dust control system ineffective. This chapter describes some of the most important, but often overlooked, measures to aid in control of workplace dust levels.

Contributing Sources/Factors

Increases in workplace dust concentrations may result from—

- Poor or inadequate maintenance of dust control systems and dust-producing equipment or processes

- Poor or inadequate housekeeping procedures

- Recirculation of previously emitted dust particulates

- Recirculation of uncontrolled emissions from dust collectors

Note: The discussion provided here assumes that adequate dust controls are installed on all major dust sources.

Administrative Measures

Number of Hours on Each Job
for Each Employee To Keep
Exposure Levels Less Than 2 mg/m³

$$\text{Cumulative Exposure} = \frac{T_1C_1 + T_2C_2 \cdots T_nC_n}{T_{total}}$$

where:

T = time

C = concentration

	Job 1 Conc. 1.0 mg/m³	Job 2 Conc. 0.5 mg/m³	Job 3 Conc. 1.2 mg/m³	Job 4 Conc. 4.5 mg/m³	Employee Cumulative Exposure mg/m³
Employee		Time Spent at Each Job			
W	0	2	4	2	1.85
X	0	4	2	2	1.67
Y	4	1	1	2	1.84
Z	4	1	1	2	1.84

Administrative measures provide a second line of defense against dust exposure. Instituting such measures also indicates management commitment to dust control. The following steps are recommended:

- Establishing a companywide dust control policy endorsed by upper management.

- Forming a dust control task force made up of plant managers, metallurgists, maintenance managers, health and safety managers, production managers, etc. The purpose of the task force is to—

  - Conduct dust surveys and identify high dust-level areas

  - Assign priorities to areas requiring dust controls

  - Suggest proper dust control techniques

  - Prepare timetables for completion of dust control projects/activities

  - Estimate costs

- Educating and training plant personnel. A short training course can be established to familiarize plant personnel with such items as—

  - Company and division dust policies

  - Government standards and safe levels

  - Reasons why dust surveys are conducted and ways to measure dust

  - Measures the company is taking to reduce dust levels and improve working conditions

  - Ways plant personnel can help

  - Types of personal protection available

- Maintaining an awareness of dust control. Information can be collected and exchanged regarding ongoing dust control programs in all divisions of the company or in related industries and government research organizations.

Periodic Dust Housekeeping Checklist

____ All mobile equipment (trucks, front-end loaders, dozers, etc.) clean.

____ Motors and switch gear clean of excessive dust, oil, slurry, and debris.

____ All operating department floors and working surfaces to be vacuumed or washed and free of scrap or debris.

____ Tools, bars, shovels, etc. stored in racks.

____ Hoses washed down on reels installed and maintained at such points as slurry pumps, stone belt transfer points, etc.

____ Safety shower operational, available, and clean.

____ Safety eyewash bottle, first-aid kit, and stretcher checked.

____ Dust accumulations around process equipment at a minimum; spills promptly cleaned up.

____ Fire extinguishers checked and available.

____ Belt-conveying systems free of spills and buildup, particularly at transfer and loading points.

____ Building siding and roofs properly maintained and free from broken or loose sheets.

____ Building doors and windows clean and in good repair at all times.

____ Locker rooms neat, clean, and in sanitary condition.

____ Offices neat and orderly.

____ Bulletin boards available and properly maintained.

____ Unpaved roadways and parking areas treated to minimize dust.

____ Yards and fields mowed frequently.

____ Plant storm sewers and open drainage ditches clean.

____ Parts inventory stored in orderly fashion.

____ Traffic lanes in shop areas clearly marked and free of scrap.

____ Abandoned process equipment and machinery promptly removed from the plant.

Preventive Maintenance Program

A preventive maintenance (PM) program is the key to reliable and efficient operation of any dust control equipment or system. Although PM programs require time and money, the investment can pay for itself through improved worker morale, increased equipment and system reliability, increased equipment and hardware life, and reduced cleanup costs and system downtime. When instituting a PM program, the following points should be considered:

- Conduct PM programs on all dust control system hardware and components, as well as dust-producing sources, during plant shutdowns or as recommended by the equipment manufacturer.

- Carry all necessary spare parts in sufficient quantities for dust control systems.

- Give high priority to patching holes, caulking and sealing cracks, and maintaining dust seals.

- Inspect and adjust all belt conveyors and their skirting rubber and dust seals.

- Inspect belt conveyor idlers and nonmoving idlers. Remove and replace missing or broken idlers.

- Inspect all belt conveyor training idlers. Adjust them as necessary so the conveyor belt does not travel laterally.

- Shut and clamp all access and inspection doors before any operation begins.

- Schedule adequate time for workers to perform routine cleanup at work stations.

- Rotate periodic cleanup among crews.

- Inspect all dust seals and repair or replace.

- Inspect belt scrapers on belt conveyors and adjust. Replace worn-out components.

- Measure velocity and static pressures weekly.

- Check for plugged ductwork and clean immediately. If this problem occurs repeatedly, redesign the ductwork.

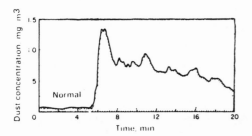

Effects of Bag Hopper Overflowing on Worker Exposure

- Develop safeguards to prevent overflowing bins.

- Follow the preventive maintenance program for dust collectors, fans, and motors described in chapter 4.

- Inspect nozzles and other components periodically. Replace worn-out nozzles as needed.

- Follow the preventive maintenance program for pumps and compressors described in chapter 3.

No dust control system can perform reliably unless it is operated properly. To achieve maximum efficiency, the following measures are suggested:

- Educate operators on startup and shutdown procedures.

- Instruct operators that all dust control systems should be in operation before any processing equipment is started.

- Eliminate the use of compressed air jets to clean accumulated dust from the equipment or clothing and substitute a vacuum cleaning system.

- Use a vacuum cleaning system to clean spills and dust accumulations. Avoid brooms and shovels.

- Use pressure water, where applicable, to clean equipment during plant shutdown or as necessary.

- Check the speed of belt conveyors and slow them down, if possible, to reduce dust circulation.

- Install an alarm to sound when a dust collector stops operating.

- Seal off obsolete working areas and remove unused equipment.

Proper Operating Procedures

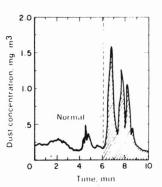

Increase in Dust Exposure When Using Compressed Air to Clean Clothing

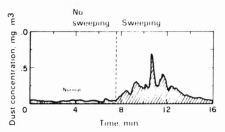

Increase in Dust Exposure When Using a Broom

Managerial Measures

A plant manager, production superintendent, or production foreman can play an important role in implementing dust control measures. The following measures can enhance the effectiveness of nonengineering controls:

● Adjust worker schedules so total daily exposure to harmful dust does not exceed threshold limit values.

● Adjust production schedule, where possible, to prevent or reduce secondary dust sources in the area. For example, empty a dust hopper at off times when no workers are in the area.

● Provide or arrange for appropriate respirators to be stored at key locations.

● Implement proper and continuous use of respirators, where necessary.

Outside Dust Source Control

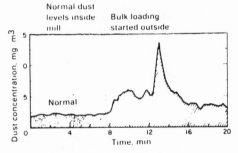

Effects of Bulk Loading Outside on Worker's Exposure Inside the Mill

Most minerals processing facilities have haul roads, loading and unloading facilities, and stockpiles. Although these dust sources are located outside the processing facility/building, dust can enter through doors, windows, air intakes, and other openings. The following measures are suggested:

● Treat the surfaces of haul roads regularly to reduce dust formation/generation due to vehicle traffic.

● Pave the haul roads, if cost permits.

● Vacuum or wash the paved roads periodically.

● Enclose all loading and unloading facilities and equip them with dust controls.

● Spray active stockpiles periodically to reduce dust spread due to wind.

● Cover and/or spray all inactive stockpiles to eliminate or reduce dust spread.

- Locate the building's air intakes in such a way to minimize outside dust infiltration.

- Use automatic bin level indicators to shut off feed to bin before overflow occurs.

A dust collection system can remove large volumes of air from a room or building. Unless this air is replaced, the room or building can come under negative pressure, or partial vacuum. Outside air brought into the building in a controlled fashion is known as make-up air.

If provisions for make-up air are not made, outside air will enter through doors, windows, and other openings in an uncontrolled fashion and may—

- Bring particulates or contaminants into the building

- "Starve" the existing dust collection system

- Create high-velocity cross drafts through doors, windows, and openings, which may be undesirable

- Affect the efficiency of natural draft stacks or chimneys

- Create unwanted drafts on workers or equipment

- Increase plant heating costs

Make-up air may be filtered and heated or cooled. In northern climates, considerable heating costs may be incurred if make-up air is heated. To reduce plant heating costs, "cleaned" exhaust gases from dust collectors may be brought into the building to serve as make-up air. This recirculation of exhaust gases can reduce plant heating costs, but it can also bring in dust if the process is not properly designed, installed, and operated.

Recirculation of Uncontrolled Emissions

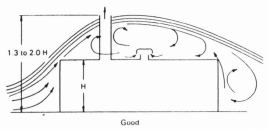

Good

High discharge stack relative to building height, air inlet on roof

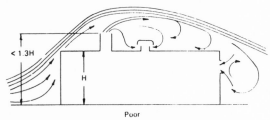

Poor

Low discharge stack relative to building height, air inlet on roof and wall

These guidelines apply only to the simple case of a low building without surrounding obstructions on reasonably level terrain

Source: American Conference of Governmental Industrial Hygienists

Building Air Inlets and Outlets

Most industrial health agencies do not recommend recirculation of exhaust gases if the dust is hazardous to health. Incorrect operation or poor maintenance—both frequent occurrences—can result in return of contaminated air into the workplace. However, under the following conditions, the recirculation of exhaust gases may be permitted.

- Installation of a dust collector or air-cleaning system of adequate efficiency so that exit concentrations do not exceed the allowable C_R values. C_R values are calculated as follows:

$$C_R = \frac{1}{2} \left[TLV - C_o \right] \frac{Q_T}{Q_R} \times \frac{1}{K}$$

where:

C_R = dust concentration in exit air (i.e., "cleaned" air) from the collector, before mixing with room air

Q_T = total air volume exhausted from the area, ft^3/min

Q_R = volume of air recirculated, ft^3/min

K = mixing factor, usually varying from 3-10, with 3 being good mixing

TLV = threshold limit value of the dust

C_o = dust concentrations measured in workers' breathing zone with exhaust gases discharged outside

- Installation of a secondary air-cleaning system of equal or greater efficiency than the primary system or provision for a reliable monitoring device to obtain and analyze a representative sample of recirculated air. Such a monitoring system should be fail-safe in the event of power failure, environmental contamination, or poor maintenance.

- Installation of a warning signal to indicate malfunction of the secondary air-cleaning system or above-limits dust concentrations.

- Provision for immediate bypass of recirculated air to the outdoors or complete shutdown of the dust-generating process when the warning signal goes off.

- Provision for proper distribution of cleaned air to allow it to mix with ambient air.

- Avoidance of high discharge velocities to prevent disturbance of the accumulated dust, which can create a local miniature dust storm.

- Recirculation of hazardous materials with low TLV's is not recommended.

Chapter 8

Sampling Dust in the Work Environment

Personal dust sampling should be a central part of the occupational health program of every mining or minerals processing operation. Fmployers have an ethical and legal duty to provide a safe, healthy working environment for their employees.

The Federal Mine Safety and Health Act was established in 1977 to protect the health and safety of miners. Section 103 of the act authorizes inspections by U.S. Government personnel; Section 101 addresses health and safety standards. Compliance with the established standards requires regular monitoring of employee exposure to harmful substances. Title 30 of the Code of Federal Regulations requires checking dust, gas, mist, and fume emissions as frequently as necessary to determine the adequacy of control measures.

In addition to ensuring compliance with the regulations, regular monitoring also provides information on—

- Existence of potential health hazards

- Possible sources and concentrations of airborne dust

- Extent of individual employee exposure to toxic substances

Personal Dust Sampling

ABC's of Personal Dust Sampling

Before conducting any personal dust sampling, a preliminary evaluation of the facility should be made. It should be conducted in two steps:

- Collect and evaluate information about the operation, such as process flowsheets describing flow of material and types of equipment used; number, type, and toxicity of raw materials, products, and by-products and the manner in which they are handled; potential sources of dust; number of workers; and types of control measures in use.

- If possible, use an instant dust monitor such as a real-time aerosol monitor (RAM) or GCA respirable dust monitor (RDM) to evaluate the existing environmental conditions. This information will save time by pinpointing dust sources, high-risk occupations/areas, etc.

Valid measurements are needed to determine if health hazards exist and dust controls are needed. The same sampling plan may not be suitable for every work and exposure environment; therefore, a sampling plan should be developed for each specific situation. Following are some of the important criteria for devising an appropriate sampling strategy:

- Type and nature of contaminant

- Location of workers and nature of work operations

- Availability of sampling equipment

- Availability of sample analytical facilities

- Availability of personnel for survey

It is not necessary to sample all workers in a facility. Suspected and potential health hazards may be evaluated by sampling a maximum risk worker—the person believed to have the greatest potential for exposure.

A worker may experience high risk because of the work area (location) or work procedures (tasks). The work area may have more than one maximum risk worker if activities or operations are not uniform or if several different exposure sources exist.

The following are important considerations when selecting the maximum risk worker:

- Proximity to contaminant sources
- Frequency of proximity to contaminant source
- Number of contaminant sources
- Worker complaints and illness

The following equation should be used to calculate TLV's for respirable dust:

$$\text{TLV respirable dust} = \frac{10 \text{ mg/m}^3}{\% \text{ respirable quartz} + 2}$$

To calculate TLV's for total dust, the following equation should be used:

$$\text{TLV total dust} = \frac{30 \text{ mg/m}^3}{\% \text{ quartz} + 3}$$

Threshold Limit Value (TLV)

Since TLV's are time-weighted based on a 7- or 8-hour work day and a 40-hour work week, MSHA uses a time-weighted average to determine compliance with the TLV's. The TWA is calculated using the following equation:

$$\frac{\text{Net dust weight (mg)}}{\text{Flow rate (L/min)} \times 0.001 \text{ (m}^3\text{/L)} \times \text{time (min)}} = \frac{\text{mg}}{\text{m}^3}$$

Short-term time periods with dust levels above the TLV are averaged with time periods below the TLV using the TWA method. When the time factor in the TWA formula is less than an 8-hour exposure, MSHA normalizes to an 8-hour exposure and calls it a shift-weighted average (SWA). This is acceptable practice provided the time interval sampled is representative of the entire shift.

Time-Weighted Average (TWA)

Personal dust samplers are used to conduct both respirable and total dust sampling. Components of a respirable dust sampler are a cyclone, a filter-cassette assembly, and a sampling pump. A total dust sampler does not have a cyclone; a filter-cassette assembly and sampling pump are its only components.

Personal Dust Sampling Equipment

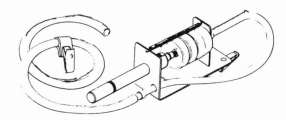

Respirable Dust Sampling Head

Total Dust Sampling Head

Cyclone

Vortex finder

Body

Grit pot

Cyclone

A cyclone is a size-selective device used to separate respirable and nonrespirable-sized particles from the air. The cyclone has the following parts:

- A vortex finder that brings the dust-laden air in at an angle and spins it.

- A cyclone body where respirable and non-respirable dust particles are separated.

- A grit pot that collects the separated nonrespirable particles.

MSHA uses a 10-mm nylon cyclone for enforcement sampling of respirable dust.

Filter-Cassette Assembly

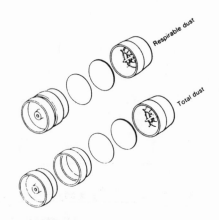

Respirable dust

Total dust

Filter-Cassette Assembly

The respirable fraction of dust that passes through the cyclone is deposited onto a filter inside a cassette. The completely assembled cassette consists of inlet and outlet plugs, top and bottom sections of the cassette, backing plate for the filter, and the filter.

MSHA specifies the following:

1. Polyvinyl chloride (PVC) membrane filters must be 37 mm in diameter and have a pore size of 5 m (for example, Millipore type PVC-5 and MSA Corporation type FWS-B).

2. A stainless steel or plastic backing screen should support the filter.

3. A two- or three-piece plastic cassette, 37 mm in diameter, should be used for respirable dust sampling while an open-face, three-piece cassette should be used for total dust sampling.

The sampling pump moves the dusty air through the sampling train. It consists of a diaphragm or piston pump driven by a battery-powered electric motor. The air volume can be controlled through a rotameter, a stroke counter, or automatically through a micro-pressure sensor. The sampling pump should operate continuously for at least 8 hours between charges.

When the respirable dust sampling head is used, the pump must be calibrated and operated at 1.7 L/min. For total dust, the pump must be recalibrated at 1.7 L/min to account for the different pressure drop of the total dust sampling head in line with the pump.

Sampling Pump

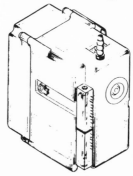

Sampling Pump

Sampling Procedure

In Office

Be sure the battery has a fresh 16-hour charge prior to each use.

Calibrate the pump using prescribed calibration procedures.

In Field

Instruct Employee

Inform the employee that the sampling train is being used to determine the amount of dust in the air and that the pump should not interfere with normal work practices.

Tell the employee when and where the sampler will be removed. Explain that if it is necessary to have the sampling device removed (for example, to use the washroom), the employee should inform a supervisor who will arrange for temporary removal. Be sure to arrange for removal of the sampling train during the lunch period.

Attach Sampling Train

Attach the pump to the employee's belt, positioned so that it does not interfere with the work operation—usually in the back on the opposite side of cassette placement. Be sure the exhaust port (charging inlet) is not obstructed. Supply a belt if the employee is not wearing one.

Position the cyclone/filter-cassette assembly on the employee's lapel or front shoulder area to approximate the breathing zone. As a guideline, attach to the left lapel of a right-handed person and the right lapel of a left-handed person. Be sure the inlet orifice is facing forward and the assembly has minimum freedom of movement. A clean, lightweight vest may be provided to secure both the pump and the filter-cassette assembly.

Initiate Sampling and Record Information

Turn on the pump. Record the starting time, pump serial number, and sample number on the sample documentation form.

Observe the pump operation for a short time to ensure that the flow rate is 1.7 L/min.

Complete the sample documentation form. Observe and record weather data or call the local weather service around midday.

Observe the pump operation after approximately 20 minutes and about every 2 hours thereafter to ensure that the sampling train is still assembled and working properly. Record the time of the check and the time of any pump flow rate adjustment in the field notes.

Collect the sample for 7 to 8 hours. If large concentrations of dust are suspected, collect two 4-hour samples and combine weights.

Conclude Sampling and Record Information

Check the flow rate and turn off the pump. Be sure to note the time that the sampling period ended.

Remove the filter cassette from the cyclone assembly. Blow loose dust from the filter cassette and **keep the assembly upright** to prevent dumping the nonrespirable-sized particles from the grit pot onto the filter.

Dust will remain in the grit pot provided the cyclone is not tilted more than 120° from vertical provided the pump is still running.

Put the inlet/outlet plugs in the filter cassette and set the sample aside. Be sure to maintain custody of all samples at all times.

Secure the collected sample. Place adhesive tape or a shrinkable cellulose band over the two-piece cassette assembly, covering the inlet and outlet plugs.

Ask the employee about activities and their duration during the sampling period to see if they match earlier predictions and if the employee classifies the day as typical.

Quality Control

Prepare a "blank," a filter-cassette assembly that is never exposed to field conditions. Remove the plugs from a prepared filter assembly in the workplace (sampling site) and immediately replace them. Seal the blank filter assembly in the same manner as a sampling filter assembly. Prepare one blank for each day of sampling. Be sure the blank filters and the filters on which the samples are collected come from the same batch since the background dust count varies from batch to batch.

Post Sampling Procedure

If a respirable crystalline silica sample was collected, each filter must be analyzed for quartz, and permissible limits must be calculated for each personal exposure. A minimum of 0.1 mg of respirable dust per filter is necessary for accurate analysis. If the presence of cristobalite or tridymite is suspected in the sample, request analysis for crystalline quartz, cristobalite, and tridymite.

Ship samples to an analytical laboratory for analysis. Be sure collected samples are shipped in a container designed to prevent damage during transit.

Errors in Respirable Dust Sampling

In minerals processing operations, compliance with the silica dust standard is based on a 1-day, full-shift respirable dust sample using a personal dust sampler. The errors associated with measuring respirable silica dust concentrations should be quantified, because the law requires a degree of certainty that the standard has been violated.

Described below are the errors associated with sampling equipment and analytical methods. Human errors are not considered here because they can be minimized or eliminated by proper care, calibration, and standard procedures.

There are four primary errors in respirable silica dust sampling:

- **Errors in Quartz Analysis** — To establish TLV's, the respirable dust sample is analyzed for quartz content. MSHA Standard Method #4 requires a precision of 11%.

- **Errors Due to Instrument Variability** — The variation between personal samplers can contribute up to 5% of the overall error factor.

- **Errors in Weighing** — A maximum of 5% error is allowable.

- **Errors Due to Fluctuations in the Pump Flow Rate** — The generally accepted value for errors in measuring airflow rate is 6%.

It is important for operators to recognize that dust sampling is a careful analytical procedure. Care should be taken to reduce errors.

The following formula for propagation of errors can be applied to calculate the overall error factor:

$$E_{overall} = \sqrt{(E_1)^2 + (E_2)^2 \ldots + (E_n)^2}$$

where:

$E_{overall}$ = error propagated by n individual errors

E_n = the nth error

Summary of Expected Errors

Source of error	Percent expected error
Quartz analysis	11
Instrument variability	5
Weighing	5
Airflow rate	6

Substituting the error values from the table above:

$$E_{overall} = \sqrt{(11)^2 + (5)^2 + (5)^2 + (6)^2}$$
$$= 14\%$$

This 14% error can be applied to the TLV's to determine the concentration above which it can be concluded that the respirable dust standard has been violated.

MSHA metal and nonmetal policy requires that the TLV be exceeded by 1.28 times for respirable and 1.10 times for total dust, before a citation is issued.

Chapter 9
Testing Dust Control Systems

Testing is necessary to evaluate the effectiveness of dust control measures. In general, testing is performed to—

- Determine if the system is operating as it was designed to operate

- Evaluate the system's effectiveness in reducing dust concentrations and employee exposure to dust

Reason for Testing

Testing a dust collection system primarily involves airflow measurement. It can provide data necessary to—

- Evaluate whether the system is performing in accordance with the design

- Set blast gates properly and adjust airflows

- Identify maintenance needs

- Determine the system's capacity for additional exhaust hoods

- Design and operate future installations effectively

Dust Collection System Testing

Test Procedure

Testing can be performed by obtaining airflow and pressure measurements at selected locations in the system. The following steps should be conducted:

- Obtain original design drawings and calculations or prepare a sketch of the system to indicate size, length, and relative location of all ducts, fittings, and associated hardware and system components. Use the drawings as a guide in selecting airflow measuring points and identifying incorrect installation or poor design.

- Measure the following:

    - Air velocity and static pressure in each branch and main

    - Static and total pressure at the fan inlet and outlet

    - Differential pressure between the inlet and outlet of the dust collector

- Analyze these measurements to find any changes from the original designs, such as a change in velocity in the branch or a change in the air volume exhausted through the hood.

Common problems that can reduce the performance of a dust collection system and their remedies are shown in the troubleshooting chart on the following page.

Airflow Measurements

Airflow inside the ductwork is usually not uniform; therefore, it is necessary to obtain velocity pressure measurements at several points within the duct cross-section. The following procedure should be used to measure velocity pressure:

- Divide the duct cross-section into equal areas. The suggested equal area grids for circular or rectangular ducts are illuatrated.

- Obtain velocity pressure measurements at the center of each of these areas.

- Convert the velocity pressure measurements into air velocities using the following relationship:

$$V = 4005 \sqrt{VP}$$

Troubleshooting Chart — Reduction in Dust Collection System Performance

Symptom	Cause	Remedy
Reduced air volume	Ducts plugged.	Clean out ducts, check air velocities, and check design specifications.
	Reduced fan speed due to belt slippage.	Check belt tension and adjust it according to manufacturer's recommendations.
	Wear or accumulation on rotor or fan casing that would obstruct airflow.	Replace or clean the worn-out equipment; consult manufacturer to see if fan is correct for the application.
	Leakage in the ductwork due to loose clean-out doors, broken joints, or worn-out ductwork.	Replace or repair leaking and worn-out sections of ductwork.
	Additional exhaust points added to the system.	Redesign or rebalance the system.
	Change of blast gates setting in branch lines.	Reset the blast gates according to original design and lock them in place.
	Increased pressure drop across dust collector.	Refer to operation and maintenance instructions of the dust collector and check its operation.

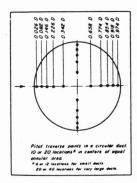

**Pitot Traverse Points
in a Circular Duct**

Reprinted by permission from the Committee on
Industrial Ventilation, Lansing, MI, 18th Edition.

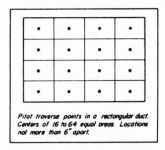

**Pitot Traverse Points
in a Rectangular Duct**

Reprinted by permission from the Committee on
Industrial Ventilation, Lansing, MI, 18th Edition.

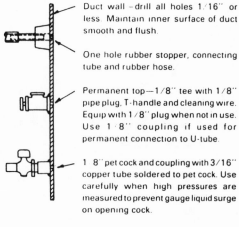

Static Tap Connections

Reprinted by permission from the Committee on
Industrial Ventilation, Lansing, MI, 18th Edition.

- Average the air velocities to obtain an average velocity (ft/min).

- Measure the inside duct diameter and calculate the cross-sectional area (ft^2).

- Multiply the duct cross-sectional area and average velocity to obtain airflow in cubic feet per minute.

Points to Note:

- The smaller the equal areas, the more accurate the measurement.

- An approved method for obtaining velocity pressure is to make two traverses, at right angles to each other, with a probe (such as a Pitot tube) across the diameter of the duct.

- Wherever possible, the traverses should be made at 8 or more duct diameters away from any major air disturbances, such as an elbow, hood, or branch entry.

- Correction for air densities should be made when air is at nonstandard conditions. For example, it is necessary to correct for air densities when—

  - Moisture content is greater than .02 lb/lb of dry air

  - Temperature of the airstream varies more than 30° F from the standard temperature

  - Altitude is greater than 1,000 ft above mean sea level

- Dust in the air may affect the instrument performance.

Static pressure measurements are made either by inserting a probe inside the ductwork or by holding a piece of tubing tightly against a small static pressure opening in the side of the duct with its other end connected to a pressure-measuring device. The characteristics of the static pressure openings in the ductwork are critical in measuring static pressure.

Points to Note:

- The static pressure openings should be—

    - Flush with the inner surfaces of the duct wall

    - Drilled and not punched

    - Without burrs or projections on the inner surface

- Pressure measurements should not be taken at the heel of an elbow or at other points where, due to excessive turbulence or change in air velocity, there is sudden expansion or contraction of ductwork.

Commonly used flow-measuring instruments are summarized in the table on the following page.

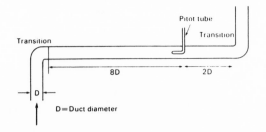

Ideal Location for Pitot Tube Measurement

Testing of wet dust suppression systems should begin with a study of the following:

- Process flow diagram

- Piping and instrumentation diagram

- Specifications for each component of the system, such as the pump, compressor, and nozzles

Wet Dust Suppression

The system should be tested in the following sequence:

1. Measure Water Flow — The flow rate of water per transfer point should be checked to verify proper operation. It is strongly recommended that direct readout flowmeters, such as rotameters, be installed near the transfer point so the flow rate can be checked quickly. If direct readout flowmeters are not installed, a circuit should be installed in the main line as shown so the flow rate can be measured easily with a portable flowmeter.

Test Procedure

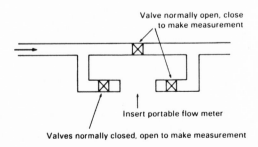

Illustration of Circuit

Flow-Measuring Instruments

Instrument	Characteristics	Advantages/Disadvantages
Pitot tube	Consists of two concentric tubes: one detects total pressure in the airstream and the other detects static pressure. Used in conjunction with monometer or Aneroid type gauge.	An extremely useful tool in flow measurement. Does not need calibration. Use in the field is limited at velocities greater than 600-800 ft/min.
U-tube manometer	Simplest type of pressure gauge. Usually filled with oil or water, but can be used with other fluids such as alcohol, mercury, or kerosene. Usually calibrated in inches of water gauge.	Can be used for either portable or stationary applications. Suitable for most static pressure measurements. Use is limited at velocities less than 1,000 ft/min (below .06 in. wg).
Inclined manometer	Improved version of the U-tube manometer. Many commercial versions equipped with a 10:1 slope, built-in level, leveling adjustment, and a means of adjusting scale to zero.	The single leg is tilted to obtain increased sensitivity and scale magnification. Accuracy dependent on slope of tube. Consequently, base must be firm enough to permit accurate leveling.
Swinging vane anemometer (Alnor velometer)	Direct readout instrument used to measure air velocities and static pressures in duct systems as well as in unrestricted areas. Consists of a meter, a measuring probe, range selectors, and connecting hoses. Velocity range is from 30 to 300 ft/min for a low-flow probe and from 100 to 10,000 ft/min for a Pitot probe. Static pressure ranges are from 0 to 1.0 in. wg and 0 to 10 in. wg.	Can be operated under high temperatures (up to 700°F) and high static pressure (up to 10 in. wg). May require periodic calibration. Presence of dust, moisture, or corrosive material can affect accuracy. A plugged filter increases resistance and the amount of air passed to the swinging vane may change. Requires a larger diameter hole than the Pitot tube (approximately 5/8 in. diameter vs. 1/4 in. to 3/8 in. for the Pitot tube).
Heated wire anemometer	Operates on principle that resistance of a wire varies with temperature and degree of temperature change is proportional to velocity of air passing over the wire. Direct readout instrument. Has short response time (less than 1 minute). Is portable. Velocity range is from 10 to 8,000 ft/min. Static pressure range is up to 10 in. wg. Temperature range is from 0 to 255°F.	Applicable for field and laboratory use. Integrity of probe must be maintained; the delicate wire can easily be damaged by mechanical shock, heavy dust loadings, or corrosive materials. Requires periodic calibration.
Aneroid type gauge	Does not use liquid. Used as field instrument in conjunction with Pitot tubes or other probes. Magnehelic gauge is the most commonly used in field installations.	Easy to read. Provides a greater response than manometers. Portable. Requires less maintenance and mounting; allows use in any position without losing accuracy. Subject to mechanical shock and failure. Requires periodic calibration.

If the water flow rate is not adequate, the following steps should be conducted:

- Check pump for—

    - Excessive vibration, noise, and heat

    - Seal or packing gland leakage

    - Discharge pressure

    - Adequate flow of mechanical seal flushing fluid

    - Lubricant supply to bearing housings

    The troubleshooting chart on the following page shows some common pump problems. For additional information, refer to the pump operations manual.

- Inspect piping for leakage at valves, tees, elbows, and drains.

- Inspect the water line filter element for excessive particle buildup.

- Inspect nozzles and replace those with physical damage.

2. Measure Compressed Airflow (where applicable) — The flow rate of air is more difficult to measure since the direct readout instruments are often calibrated for a certain pressure. The flow rate for different pressures can be calculated using manufacturers' charts, actual pressure, and the flow rate at a calibrated pressure. If a flowmeter is not installed in the air line, a circuit as shown for testing water flow rate should be installed.

If the compressed airflow and pressures are not adequate, the following items should be checked:

- Check air compressor for—

    - Correct adjustment of upper and lower pressure set points

    - Excessive noise and vibration

    - Leakage in the line

Troubleshooting Chart – Pump Problems

Symptom	Cause	Solution
No liquid delivered.	Lack of priming.	Fill pump and suction pipe completely with liquid.
	Suction lift too high.	If no obstruction at inlet, check for pipe friction losses. If static lift is too high, liquid to be pumped must be raised or pump lowered.
	Discharge head too high.	Check pipe friction losses. Check that valves are wide open.
	Impeller plugged.	Dismantle pump and clean impeller.
Not enough liquid delivered.	Air leaks in suction piping.	Test flanges for leakage. Suction line can be tested by plugging inlet and putting line under pressure.
	Impeller partially plugged.	Dismantle pump and clean impeller.
	Defective impeller.	Inspect impeller and shaft. Replace if damaged or vane sections badly eroded.
	Defective packing or seal.	Replace packing or mechanical seal.
	Suction pipe not immersed enough.	Lower inlet pipe.
Not enough pressure.	Speed too low.	Check whether motor is receiving full voltage.
	Air leaks in suction piping.	Test flanges for leakage. Suction line can be tested by plugging inlet and putting line under pressure.
	Mechanical defects.	Inspect impeller and shaft. Replace if damaged or vane sections badly eroded. Replace packing or mechanical seal.
	Obstruction in liquid passages.	Dismantle pump, inspect passages, and remove obstruction.
	Air or gases in liquid. (Watch for bubbles.)	Possibility of overrated pump. Periodically exhaust accumulated air.
Pump operates for short time, then stops.	Incomplete priming.	Free pump, piping, and valves of all air. Correct any high points in suction line.
	Air leaks in suction piping.	Test flanges for leakage. Suction line can be tested by plugging inlet and putting line under pressure.
	Air leaks in stuffing box.	Increase seal liquid pressure to above atmosphere.
Pump takes too much power.	Mechanical defects.	Inspect impeller and shaft. Replace if damaged or vane sections badly eroded. Replace packing or mechanical seal.
	Suction pipe not immersed enough.	Lower inlet pipe.
	Shaft bent or damaged.	Check deflection of rotor by turning on bearing journals.
	Failure of pump parts.	Check bearings and impeller for damage.

- Proper lubrication

Refer to the operations manual for additional troubleshooting of the air compressor.

- Inspect piping for leakage at valves, tees, elbows, and drains.

- Drain the compressed air line filters.

- Replace depleted compressed air dryer element.

3. Check unique components such as metering units, foaming units, and electrostatic charge generators for proper operation.

4. During freezing weather conditions, check for proper operation of heat tracing elements or tapes and insulation to avoid freezing in pipes.

Evaluating Dust Control Systems

After a dust control system is installed, it is essential to evaluate the system's effectiveness in reducing dust concentrations and associated employee exposure to dust. A sampling plan must be developed before any evaluation can be made.

Types of Samples

Two primary types of samples are normally obtained:

- Process/source samples to measure airborne dust concentrations due directly to the source emissions

- Ambient/background samples to measure the contribution of other sources to dust levels

Sampling Locations

The selection of sampling locations is important because of the large differences in dust levels that may exist around a dust source. Many times, a change in the sampling location can have a greater effect on dust concentrations measured than the effect of engineering controls being evaluated.

Process or source samples should be located close to the source to reduce the interference of other sources in the vicinity. They may also be located near the worker whose exposure is most directly affected by the dust source.

Ambient/background samples should be located far enough so as not to be affected by dust emissions from the source, yet close enough to be representative of dust levels generated by all other sources in the vicinity.

Sampling Instrumentation

Two sets of sampling instruments are commonly used—

* Instantaneous dust monitors, such as RAM-1, to obtain real-time dust concentrations

* Gravimetric samplers to provide time-weighted average dust concentration data and mineral composition analysis

Instant dust monitors provide immediate dust concentration information. They may not be compared with the 8-hour time-weighted gravimetric samplers used for compliance monitoring; however, they are useful for rapid identification of major dust sources as well as rapid evaluation of dust control measures.

Gravimetric dust samplers provide a time-weighted average of dust concentration for the entire sampling period. They do not provide the information required to pinpoint sources of dust or determine how the concentrations vary as a function of operating conditions. Gravimetric dust sampler results should be used with caution.

Sampling Approaches

The three most common sampling approaches available are—

1. Obtaining short-term samples with system "on" and "off."

2. Obtaining samples before and after installation of a dust control system.

3. Obtaining samples to compare the effectiveness of two different types of dust control systems.

Approaches 1 and 2 provide data to determine the effectiveness of a newly installed dust control system. In approach 1, the samples are obtained after the system is installed; in approach 2, the samples are obtained before and after the system is installed. Approach 3, commonly known as the A-B-A approach, is useful when it is essential to compare the effectiveness of two dust control systems during the same test period. It involves a period of testing with system A, followed immediately by an equivalent period of testing with system B, followed immediately by verification of the performance of system A. The return to system A provides an indication that changes observed with system B were not due to other changes in the process.

Data Collection

A data collection form should be developed to record all relevant information during field testing.

Strip chart recorders or more recent electronic data logging units can be attached to the instantaneous dust monitors to provide records and sound indications of dust levels. Small, portable tape recorders are often used, especially with instantaneous dust monitors, to record the data, field observation, and any other significant activity connected with the testing. When tape recorders are used, the data should be transcribed every day onto the appropriate data sheets. This will facilitate review of the data on a daily basis and prevent any future problems or omissions. It will also provide a preliminary feel for the system's effectiveness and indicate whether the testing program is going as planned.

Data Analysis

To evaluate a system's effectiveness in reducing dust emissions, two methods are used to analyze the dust concentration data:

- The graphical method
- The mathematical method

The graphical method, illustrated on the following page, is a qualitative method. It provides a simple, yet effective, method to compare the dust concentration data with system "off" and system "on" and determines the magnitude of the system's effectiveness in reducing dust concentrations.

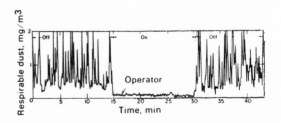

**ABA Comparison of
Dust Control Method**

Mathematical methods are used to quantify the efficiency of system performance. Here, the system "off" and system "on" dust concentration data are compared and the system's efficiency is determined using the following equation:

$$\eta = \frac{C_{off} - C_{on}}{C_{off}} \times 100\%$$

where:

η = system's efficiency, in percent

C_{off} = system "off" dust concentrations, mg/m^3

C_{on} = system "on" dust concentrations, mg/m^3

If sampling is repeated for a number of times at the same location, then the data should be treated statistically for more detailed analysis. Information on statistical approaches can be obtained through several excellent reference books.

Bibliography

Adkins, J.H., P. Krois, M. Hinton, and J. Nussbaumer. Baseline Training Materials for Ass. Compliance and Acc. Red. in the Metal and Nonmetal Mining Industries, 1983.

Alden, J.L., and J.M. Kane. Design of Industrial Ventilation Systems, 1982.

Allen, R.W. Effect of Particle Wettability on Droplet Target Efficiencies in Wet Scrubbers, 1976.

Anderson, D.M. "Dust Control Design by the Air Induction Technique," Int. Medicine and Surgery, February 1964, pp. 68-72.

Apt, J., and F.G. Anderson. Respirable Dust Control Manual for Underground Coal Mines, Vol. 2, U.S. Bureau of Mines publication, Apt, Bramer, Conrad and Associates, Contract HO111464, NTIS-PB-219616, 1972.

Armbruster, L., and H. Breuer. "Dust Monitoring and the Principle of On-Line Dust Control," International Symposium on Aerosols in the Mining and Industrial Work Environment.

Armbruster, L., H. Breuer, D. Mark, and J.H. Vincent. "The Definition and Measurement of Inhalable Dust," International Symposium on Aerosols in the Mining and Industrial Work Environment, 1981.

Balakrishnan, N.S., G.H. Cheng, and M.N. Patel. Emerging Technologies for Air Pollution Control, 1979.

Bartley, D.L., and G.M. Breuer. "Analysis and Optimization of the Performance of the 10mm Cyclone," American Industrial Hygiene Association Journal, July 1982, pp. 520-528.

Bauer, H.D. "Dedusting Exp. at a Dump Where Coal Is Loaded From a Face Into the Roadway Conveyor," Proceedings of Conference on Technical Measurement of Dust Prevention and Suppression in Mines, Luxembourg, Belgium, October 11-13, 1972, pp. 533-546.

189

Baumeigter, T., E.A. Avallone, and T. Baumeigter III. Marke Standard Handbook for Mechanical Engineers, 1978.

Becker, H. "Tasks and Trends in Dust Control in German Coalmines," Symposium on Environmental Engineering in Coal Mining, Harrogate, October 31-November 2, 1972, pp. 135-140.

Bender, M. "Fume Hoods, Open Canopy Type—Their Ability To Capture Pollutants in Various Environments," American Industrial Hygiene Association Journal, February 1979, pp. 118-127.

Brief, R.S. Basic Industrial Hygiene: A Training Manual, American Industrial Hygiene Association, Winter Supplement to the Medical Bulletin of Exxon Corporation, 1975.

Bryant, R.K. "Ductwork—How To Make the System Work," Proceedings of 22nd Annual Industrial Ventilation Conference, March 1980.

Burke, E. "Dust Control in the Cement Industry," Cement Lime and Gravel, November 1969, pp. 317-323.

Canton, A.T., W.L. Evans, and C.R. Smith. "Selection of Air Pollution Control Plant for the Minerals Processing Industry," Filtration and Separation, January/February 1976, pp. 32-45.

Cecala, A.B., J.C. Volkwein, and E.D. Thimons. New Bag Nozzle To Reduce Dust From Fluidized Air Bag Machines, 1984.

Cecala, A.B., J.C. Volkwein, E.D. Thimons, and C.W. Urban. Protection Factors of the Airstream Helmet, U.S. Bureau of Mines Report of Investigations 8591, 1981.

Cheng, L. "Collection of Airborne Dust by Water Sprays," Ind. Eng. Chem. Proc. Develop., Vol. 12, No. 3, 1973, pp. 221-225.

Cheng, L. "Formation of Airborne-Respirable Dust at Belt Conveyor Transfer Points," American Industrial Hygiene Association Journal, Vol. 34, No. 12, December 1973, pp. 540-546.

Cheng, L. "Dynamic Spreading of Drops Impacting Onto a Solid Surface," Ind. Eng. Chem. Proc. Develop., Vol. 16, No. 2, 1977, pp. 192-197.

Cheng, L. "Optimizing Water Sprays for Dust Suppression," Eng. Min. Journal, Vol. 179, No. 10, October 1978, pp. 115-118.

Cheremisinoff, P.N., and N.P. Cheremisinoff. "Fabric Filters for Dust Collection—Part 2, Types of Fabrics," Plant Engineering, June 1973, pp. 152-155.

Cheremisinoff, P.N., and N.P. Cheremisinoff. "Fabric Filters for Dust Collection—Part 1, What They Are and How They Work," Plant Engineering, May 1973, pp. 62-65.

Cheremisinoff, R.N., and R.A. Young. Pollution Engineering Practice Handbook, 1976.

Clarke, P.D., and P. Swift. "The Development of Unit Solutions to Industrial Dust Control Problems," Filtration and Separation, May/June 1978, pp. 242-252.

Clauser, R. "A Simple and Inexpensive Hood for Dust Control," Chemical Engineering, July 1975, p. 106.

Cole, H.W. "Foam Suppressants for Control of Dust," Coal Mining and Processing, Vol. 14, No. 10, October 1977, pp. 98-104.

Colijn, H., and P.J. Conners. "Belt Conveyor Transfer Points," Trans. Soc. Min. Eng., June 1972, pp. 204-252.

Committee on Industrial Ventilation, American Conference of Governmental Industrial Hygienists. Industrial Ventilation, 18th Edition. 1984.

Conway, J.R., and R.J. Synk. A Business Look at Baghouse Specifications, Vol. 29, No. 1, 1979.

Courtney, W.C., N.I. Jayaraman, and P.C. Behum. Effect of Water Sprays for Respirable Dust Suppression With a Research Continuous Mining Machine, U.S. Bureau of Mines Report of Investigations 8283, 1978.

Courtney, W.G. "Single Spray Reduces Dust 90%," Coal Mining and Processing, June 1983, pp. 75-77.

Courtney, W.G., and L. Cheng. "Control of Respirable Dust by Improved Water Sprays."

Cross, F.L. "Choosing a Liner for Better Venturi Scrubber Maintenance," Pollution Engineering, March 1978, pp. 52-53.

Cross, F.L. Control of Fugitive Dust From Bulk Loading Facilities, 1980.

Cross, F.L., and H.E. Hesketh. Handbook for the Operation and Maintenance of Air Pollution Control Equipment, 1975.

Dalla Valle, J.M. Exhaust Hoods, The Industrial Press, Inc., New York, New York, 1952.

Danielon, J.A. Air Pollution Engineering Manual, 2nd ed., Air Pollution Control District, County of Los Angeles, AP-40, 1973.

Dennis, R. "Approaches to Dust Suppression," GCA Corporation, personal communication, 1981.

Dennis, R., and D.V. Bubenick. Fugitive Emissions Control for Solid Materials Handling Operations, 1983.

Dickie, L. "Unit Collectors Versus Central Systems," Heating/Piping/Air Conditioning, March 1977.

Dickie, L. "Controlling Airborne Dust on Conveyor Belt Systems," Coal Mining and and Processing, Vol. 15, No. 1, 1978, pp. 72-74.

Dickie, L.   "Insertable Dust Collectors Clean Up Material Handling," Pit and Quarry, March 1978, pp. 62-63.

Dumbaugh, G.D.   "Flow From Bulk-Solids Storage Bins," Chemical Engineering, March 1979, pp. 189-193.

Ebly, R.W., and R.H. Uttke.   "Conveyor Idler Maintenance and Selection Are the Keys to Efficient Operation," Rock Products, March 1978, pp. 66-69.

Emory, S.F., and J.C. Berg.   "Surface Tension Effects on Particle Collection Efficiency," Appendix to U.S. Environmental Protection Agency Report No. EPA-600/7-78-097, 1978.

Evans, D.   "Ambitious Dust Control Program Makes Cassiar a Better Place To Work," Canadian Mining Journal, November 1977, pp. 24-29.

Evans, R.J.   Methods and Costs of Dust Control in Stone Crushing Operations, U.S. Bureau of Mines Info. Circular 8669, 1975.

First, M.W.   "Engineering Control of Occupational Health Hazards," American Industrial Hygiene Association Journal, September 1983, pp. 621-626.

Folwell, J.   "Design of Hoods for Dust Control Systems," Proceedings of Dust Control Symposium, Institute of Chemical Engineers and Institute of Materials Handling, University of Salford, U.K., March 21-22, 1978, pp. 1-29.

Ford, V.H.   "Bottom Belt Sprays as a Method of Dust Control on Conveyors," Mining Technology, September 1973, pp. 387-391.

Gibson, N.   "Design of Explosion Protection for Dust Control Equipment," Dust Control Symposium, 1978.

Goddard, B., K. Bower, and D. Mitchell.   Control of Harmful Dust in Coal Mines, 1980.

Goodfellow, H.D., and M. Bender.   "Design Considerations for Fume Hoods for Process Plants," American Industrial Hygiene Association Journal, July 1980, pp. 473-484.

Goodfellow, H.D., and M. Bender.   "Environmental Design Considerations for Belt Conveyors," CIM Bulletin, September 1982, pp. 97-104.

Gorse, D.E., and W. Scott.   "Air Slides Conveyance for Better Dust Control," 46th Annual Technical Session, Mines Accident Prevention Association of Ontario, 1977.

Graham, K.B.   "Industrial Fan Selection and Installation," Canadian Mining Journal, October 1975, pp. 69-72.

Grassmuck, G.   "Dust Control in Drilling and Blasting," Canadian Pit and Quarry, August 1967, pp. 26-28.

Grover, S.N., H.R. Pruppacher, and A.E. Hamielec. "A Numerical Determination of the Efficiency With Which Spherical Aerosol Particles With Spherical Water Drops, Due to Inert. Impact., Phoretic, and Elec. Forces," J. Atmos. Environ., Vol. 34, October 1977, pp. 1655-1663.

Haemon, W.C. Plant and Process Ventilation, 2nd ed., Industrial Press, Inc., New York, New York, 1963.

Hagopian, R.J., and E.K. Bastress. Engineering Control Research Recommendations, NIOSH Technical Information, NTIS-PB-273798, February 1976.

Hama, G.M. "Ventilation for Control of Dust From Bagging Operations," Heating and Ventilating, April 1973.

Hama, G.M. "Jet Stream Replaces Vertical Duct in Push-Pull Ventilation of Pickling Tanks," Heating/Piping/Air Conditioning, January 1973.

Harrold, R. "Surfactants vs. Dust—Do They Work?" Coal Age, Vol. 84, No. 6, June 1979, pp. 102-105.

Hatch, T. "Dust Control: Present and Future Design Considerations," Mechanical Engineering, Vol. 57, 1935, pp. 154-156.

Hatch, T. "Fundamental Factors in the Design of Exhaust Systems," Mechanical Engineering, February 1935, pp. 109-113.

Hatch, T. "Design of Exhaust Hoods for Dust-Control Systems," Journal of Industrial Hygiene and Toxicology, November 1936, pp. 595-603.

Hazard, W.G. "Exhaust and Ventilation—Part 2: Specific Applications," National Safety News, December 1978, pp. 81-85.

Hazard, W.G. "Exhaust and Ventilation—Part 1: Basic Principles," National Safety News, November 1978, pp. 76-83.

Hiltz, R.H. Underground Application of Foam for Suppression of Respirable Dust, U.S. Bureau of Mines publication, MSA Research Corporation, NTIS-PB-249861, September 1975.

Hiltz, R.H., and J.V. Friel. "Using High Expansion Foam To Control Respirable Dust," Min. Congr. J., Vol. 59, No. 5, May 1973, pp. 54-60.

Hodgson, J.M. "Dust Control in Quarries: Screening Operations," Quarry Manager's Journal, Vol. 5, June 1967, pp. 231-232.

Hoenig, S.A. "Application of Electrostatically Charged Fog to the Suppression of Respirable Dust," Pit and Quarry, August 1976, pp. 88-90.

Hoenig, S.A. Use of Electrostatically Charged Fog for Control of Fugitive Dust Emissions, EPA-600/7-77-131, November 1977.

Hoenig, S.A. Electrostatic Charging of Dust and the Control of Industrial Dust, Fume and Smoke by Charged Water Fog, 1978.

Kinnear, D.I., J. Edwards, and I.J. Webb. "On Site Testing of High Efficiency Air Filtration Installations," Dust Control Symposium, 1978.

Kleysteuber, W. Safety Evaluation of Conveyor Belt Cleaning Systems, U.S. Bureau of Mines, January 1983.

Kost, J.A., G.A. Shirey, and C.T. Ford. In-Mine Test for Wetting Agent Effectiveness, U.S. Bureau of Mines publication, NTIS-PB82-183344, 1980.

Kost, J.A., J.C. Yingling, and B.J. Mondics. Guidebook for Dust Control in Underground Mining, U.S. Bureau of Mines publication, Bituminous Coal Research, Inc., Contract JO199046, NTIS-PB83-109207, 1981.

Kruse, C.W., and W.O. Bianconi. "Air Flow Induced in Enclosed Inclined Chutes of Material Handling Systems," American Industrial Hygiene Association Journal, May-June 1966, pp. 220-227.

Larson, S.L. "Air Induction by Falling Materials as a Basis for Exhaust Hood Design," M.S. Thesis, University of Pittsburgh, 1952.

Lenox, A.E. "A Practical Look at Dust Control," Heating/Piping/Air Conditioning, November 1977.

Mathai, C.V., and L.A. Rathbun. "An Electrostatically Charged Fog Generator for the Control of Inhalable Particles," Proceedings EPA 3rd Symposium on the Transfer and Utilization of Particulate Control Technology, Orlando, Florida, March 9-12, 1981.

Mathai, C.V., L.A. Rathbun, and D.C. Drehmel. "Prototype Tests of a Charged Water Droplet Generator for the Control of Inhalable Fugitive Dust," 74th Annual Meeting of the Air Pollution Control Association, Philadelphia, Pennsylvania, June 1981.

McCaffery, R.A. "Techniques of Effective Dust Control in the Mining and Metallurgy Industry," Annual General Meeting of the Canadian Institute of Mining and Metallurgy, Quebec, April 1971.

McCaffray, N. "Do's and Don'ts," personal communication to Vinit Mody, 1979.

McCoy, J., J. Melcher, J. Valentine, D. Monaghan, T. Muldoon, and J. Kelly. Evaluation of Charged Water Sprays for Dust Control, U.S. Bureau of Mines publication, NTIS-PB83-210476, January 1983.

Mining Association of Canada. Design Guidelines for Dust Control at Mine Shaft and Surface Operations, 3rd edition, April 1980.

Mody, V., and R. Jakhete. Conveyor Belt Dust Control, U.S. Bureau of Mines publication, Contract H0113007, July 1984.

Moechnig, B.   "Dust Control in Enclosed Conveyors Using Water Fog," personal communication, Cargill, Inc., 1979.

Morrison, J.N.   "Controlling Dust Emissions at Belt Conveyor Transfer Points," Trans. Soc. Min. Eng., Vol. 250, AIME, March 1971, pp. 47-53.

Murphy, A.S.   "Dust Control at the Primary Crusher/ Loadout Level," Canadian Mining Journal, October 1971, pp. 72-73.

Myers, R.C.   "Industrial Fans—Guidelines for a Successful Installation," Iron and Steel Engineer, October 1976, pp. 38-44.

National Research Council.   Measurement and Control of Respirable Dust in Mines, NMAB-363, National Academy of Science, Washington, D.C., 1980.

Neveril, R.B., J.U. Price, and K.L. Engdahl.   Capital and Operating Costs of Selected Air Pollution Control Systems—I, 1978.

Neveril, R.B., J.U. Price, and K.L. Engdahl.   Capital and Operating Costs of Selected Air Pollution Control Systems—III, 1978.

Nilsson, I.L.   "The Design of Dust Enclosures by Means of the Trellex Dust Sealing System," Proceedings of Dust Control Symposium, Institute of Chemical Engineers and Institute of Materials Handling, University of Salford, U.K., March 21-22, 1978, pp. 2-1 to 2-6.

Page, S.J.   An Evaluation of Three Wet Dust Control Techniques for Face Drills, U.S. Bureau of Mines Report of Investigations 8596, 1982.

Page, S.J.   Evaluation of the Use of Foam for Dust Control on Face Drills and Crushers, U.S. Bureau of Mines Report of Investigations 8595, 1982.

Pereira, N.C., and L.K. Wang.   "Air and Noise Pollution Control," Handbook of Environmental Engineering, Volume One, Humana Press, Clifton, New Jersey, 1979.

Powlesland, J.W.   "New Uses Continue To Be Found for Air Curtains," Canadian Mining Journal, October 1971, pp. 84-93.

Pring, R.T.   "Dust Control in Large-Scale Ore-Concentrating Operations," Mining Technology, Amer. Inst. of Min. and Met. Engs., Tech. Pub. No. 1225, September 1940.

Pring, R.T., J.F. Knudsen, and R. Dennis.   "Design of Exhaust Ventilation for Solid Materials Handling: Fundamental Considerations," Ind. Eng. Chem., Vol. 41, No. 11, pp. 2442-2450.

Rankin, R.L., S.J. Rodgers, and E.V. Polite.   Effects of Engineering Parameters on the Control of Respirable Dust, U.S. Bureau of Mines publication, NTIS-PB83-232207, November 1982.

Reisinger, A.A. "Fabric Filter Collectors—Proper Maintenance Procedures," Pit and Quarry, October 1975, pp. 89-92.

Ricketts, D.B., and R.W. Bock. "Dust Collection—Material Handling Operations, Sinter Plants," Ironmaking Proc. Metall. Soc., Vol. 36, AIME Iron Steel Division, Pittsburgh, Pennsylvania, April 17-20, 1977.

Rodgers, S.J., R.L. Rankin, and M.D. Marshall. Improved Dust Control at Chutes, Dumps, Transfer Points, and Crushers in Noncoal Mining Operations, U.S. Bureau of Mines publication, NTIS-PB-297422, 1978.

Rosen, K.M. "Practical Ducting Design, Part 11—Calculating Duct Losses," Plant Engineering, Vol. 26, No. 22, November 1972, pp. 55-58.

Rosenberger, F. "Fully Automated Train Loading of Limestone Products," Pit and Quarry, November 1975, pp. 76-78.

Ross, C.R. Prevention and Suppression of Dust in Mining, Tunnelling and Quarrying in Canada, 1958-1962, Occupational Health Division, Department of National Health and Welfare, Ottawa, 1963.

Ross, J.L. "Chemical Eliminates Haulroad Dust," Coal Age, June 1977.

Schlick, D.P. Respirable Dust Control in the Mines of West Germany, U.S. Bureau of Mines Info. Circular 8490.

Schofield, C., H.M. Sutton, and K.A. Waters. "Generation of Dust by Materials Handling Operations," Proceedings of Dust Control Symposium, Institute of Chemical Engineers and Institute of Materials Handling, University of Salford, U.K., March 21-22, 1978.

Schowengerdt, F.D., and J.T. Brown. "Colorado School of Mines Tackles Control of Respirable Coal Dust," Coal Age, April 1976.

Seibel, R.J. Dust Control at a Transfer Point Using Foam and Water Sprays, U.S. Bureau of Mines Respirable Dust Program Technical Progress Report 97, May 1976.

Soderberg, H.E. "How To Select Ductwork, Hoods and Dust Collectors," Engineering and Mining Journal, Vol. 161, June 1960.

Soderberg, H.E. "Dust Control in Preparation Facilities," American Air Filter Com., presented at the 1978 Coal Convention of the American Mining Congress, 1978.

Stahura, R. "Making Conveyor-Belt Cleaners Work," Plant Engineering, August 1978, pp. 88-90.

Stern, A.C., K.J. Caplean, and P.D. Bush. Cyclone Dust Collectors, 1956.

Taylor, D.H. "Recommendations for Dust Collection Systems," Metal Progress, December 1970, p. 63.

Tomb, T.F., and M. Corn. Comparison of Equiv. Spherical Volume and Aerodynamic Diameters for Irregularly Shaped Particles, U.S. Bureau of Mines publication, 1972.

U.S. Environmental Protection Agency. Air Pollution Engineering Manual, Publication AP-40, May 1973.

Valle, D.O. Chapter 1, "Theory of Flow of Gases Into an Opening," and Chapter 11, "Hood Entrance Losses," Exhaust Hoods.

Valliere, H.P. "Cement Plant Dust Control," Annual General Meeting of the Canadian Institute of Mining and Metallurgy, Quebec, April 1971.

Volkwein, J.C., A.B. Cecala, and E.D. Thimons. Adding Foam to Dust Control in Minerals Processing, Pittsburgh Research Center, U.S. Bureau of Mines.

Volkwein, J.C., A.B. Cecala, and E.D. Thimons. Use of Foam for Dust Control in Minerals Processing, U.S. Bureau of Mines Report of Investigations 8603, 1983.

Volkwein, J.C., A.B. Cecala, and E.D. Thimons. Effectiveness of Three Water Spray Methods Used To Control Dust During Bagging, U.S. Bureau of Mines Report of Investigations.

Walker, C.L. "A Guide to Dust Collection Planning," Pit and Quarry, November 1979, pp. 95-97, 104.

Walli, R.A. "Dust Collection With Fabric Filters: Some Theory and Applications in the Mineral Industries," Annual General Meeting of the Canadian Institute of Mining and Metallurgy, Quebec, April 25-28, 1971.

Walton, W.H., and A. Woolcock. "The Suppression of Airborne Dust by Water Spray," Aerodynamic Capture of Particles, Pergamon Press, London, 1960, pp. 129-153.

Warrington, G.B. "A New Dust Abatement in a Crushing and Screening Plant," Proceedings 22nd Annual Meeting Aggregate Producers Assoc. of Ontario, Ottawa, March 2, 1979.

Woffinden, G.J., Q.R. Markowski, and D.S. Ensor. "Effects of Surface Tension on Particle Removal," Symposium on the Transfer and Utilization of Particulate Control Technology, Vol. 3.

Wright, R.D. "Design and Calculation of Exhaust Systems for Conveyor Belts, Screens and Crushers," Journal of the Mine Ventilation Society of South Africa, January 1966, pp. 1-7.

Yourt, G.R. Design Guidelines for Dust Control at Mine Shafts and Surface Operations, 1980.

Index

RADON
AND THE ENVIRONMENT

Edited by

William J. Makofske and Michael R. Edelstein

Institute for Environmental Studies
Ramapo College of New Jersey

This volume provides an interdisciplinary overview and analysis of radon and the environment, geared to both professional and lay perspectives. It is based on a conference of the same title sponsored by the Institute for Environmental Studies.

The radon issue spans many disciplines and has far-reaching implications for society. There are also many uncertainties stemming from a variety of sources. These include the often misleading and inconsistent media coverage of the topic, the newness of the issue, the lack of detailed scientific information and the way people perceive and respond to risk. While the effects of radon are still not fully understood as a public policy and health issue, there have been important new developments on the subject; and this book brings together many of the key contributors to our current knowledge. It attempts to clarify the policy issues, in a manner that will be of equal use to a radon professional, a government official, or a concerned citizen.

Seven aspects of the radon issue are presented in the various sections of the book. The condensed table of contents given below lists **section titles and selected chapter titles.**

ISBN 0-8155-1161-2 (1988)

465 pages

Other Noyes Publications

MINIMIZING EMPLOYEE EXPOSURE TO TOXIC CHEMICAL RELEASES

by

**Ralph W. Plummer, Terrence J. Stobbe,
James E. Mogensen, Luanne K. Jeram**
West Virginia University

Pollution Technology Review No. 145

This book describes procedures for minimizing employee exposure to toxic chemical releases and suggested personal protective equipment (PPE) to be used in the event of such chemical release. How individuals, employees, supervisors, or companies perceive the risks of chemical exposure (risk meaning both probability of exposure and effect of exposure) determines to a great extent what precautions are taken to avoid risk.

The objective of this study was to obtain information from chemical manufacturers and facilities which use chemicals in other processes, to determine what types of procedures are currently in practice, and to develop a set of recommended procedures and suggested PPE which can be used to minimize the possibility and the risk of toxic chemical releases. In Part I, the authors develop an approach which divides the project into three phases: kinds of procedures currently being used; the types of toxic chemical release accidents and injuries that occur; and, finally, integration of this information into a set of recommended procedures which should decrease the likelihood of a toxic chemical release and, if one does occur, will minimize the exposure and its severity to employees. Part II covers the use of personal protective equipment. It addresses the questions: what personal protective equipment ensembles *are used* in industry in situations where the release of a toxic or dangerous chemical may occur or has occurred; and what personal protective equipment ensembles *should be used* in these situations.

A condensed table of contents listing **part titles, chapter titles and selected subtitles** is given below.

ISBN 0-8155-1131-0 (1987)

257 pages